AN INTRODUCTION TO MATHEMATICS

Alfred North Whitehead

ILLUSTRATED
&
PUBLISHED BY
E-KİTAP PROJESİ & CHEAPEST BOOKS

www.cheapestboooks.com

www.facebook.com/EKitapProjesi

ISBN: 978-625-6629-62-2

CONTENTS

AN INTRODUCTION TO MATHEMATICS

CHAPTER I

THE ABSTRACT NATURE OF MATHEMATICS

THE study of mathematics is apt to commence in disappointment. The important applications of the science, the theoretical interest of its ideas, and the logical rigour of its methods, all generate the expectation of a speedy introduction to processes of interest. We are told that by its aid the stars are weighed and the billions of molecules in a drop of water are counted. Yet, like the ghost of Hamlet's father, this great science eludes the efforts of our mental weapons to grasp it—" 'Tis here, 'tis there, 'tis gone"—and what we do see does not suggest the same excuse for illusiveness as sufficed for the ghost, that it is too noble for our gross methods. "A show of violence," if ever excusable, may surely be "offered" to the trivial results which occupy the pages of some elementary mathematical treatises.

The reason for this failure of the science to live up to its reputation is that its fundamental ideas are not explained to the student disentangled from the technical procedure which has been invented to facilitate their exact presentation in particular instances. Accordingly, the unfortunate learner finds himself struggling to acquire a knowledge of a mass of details which are not illuminated by any general conception. Without a doubt, technical facility is a first requisite for valuable mental activity: we shall fail to appreciate the rhythm of Milton, or the passion of Shelley, so long as we find it necessary to spell the words and are not quite certain of the forms of the individual letters. In this sense there is no royal road to learning. But it is equally an error to confine attention to technical processes, excluding consideration of general ideas. Here lies the road to pedantry.

The object of the following Chapters is not to teach mathematics, but to enable students from the very beginning of their course to know what the science is about, and why it is necessarily the foundation of exact thought as applied to natural phenomena. All allusion in what follows to detailed deductions in any part of the science will be inserted merely for the purpose of example, and care will be taken to make the general argument comprehensible, even if here and there some technical process or symbol which the reader does not understand is cited for the purpose of illustration.

The first acquaintance which most people have with mathematics is through arithmetic. That two and two make four is usually taken as the type of a simple mathematical proposition which everyone will have heard of. Arithmetic, therefore, will be a good subject to consider in order to discover, if possible, the most obvious characteristic of the science. Now, the first noticeable fact about arithmetic is that it applies to everything, to tastes and to sounds, to apples and to angels, to the ideas of the mind and to the bones of the body. The nature of the things is perfectly indifferent, of all things it is true that two and two make four. Thus we write down as the leading characteristic of mathematics that it deals with properties and ideas which are applicable to things just because they are things, and apart from any particular feelings, or emotions, or sensations, in any way connected with them. This is what is meant by calling mathematics an abstract science.

The result which we have reached deserves attention. It is natural to think that an abstract science cannot be of much importance in the affairs of human life, because it has omitted from its consideration everything of real interest. It will be remembered that Swift, in his description of Gulliver's voyage to Laputa, is of two minds on this point. He describes the mathematicians of that country as silly and useless dreamers, whose attention has to be awakened by flappers. Also, the mathematical tailor measures his height by a quadrant, and deduces his other dimensions by a rule and compasses, producing a suit of very ill-fitting clothes. On the other hand, the mathematicians of Laputa, by their marvellous invention of the magnetic island floating in the air, ruled the country and maintained their ascendency over their subjects. Swift, indeed, lived at a time peculiarly unsuited for gibes at contemporary mathematicians. Newton's *Principia* had just been written, one of the great forces which have transformed the modern world. Swift might just as well have laughed at an earthquake.

But a mere list of the achievements of mathematics is an unsatisfactory way of arriving at an idea of its importance.

It is worth while to spend a little thought in getting at the root reason why mathematics, because of its very abstractness, must always remain one of the most important topics for thought. Let us try to make clear to ourselves why explanations of the order of events necessarily tend to become mathematical.

Consider how all events are interconnected. When we see the lightning, we listen for the thunder; when we hear the wind, we look for the waves on the sea; in the chill autumn, the leaves fall. Everywhere order reigns, so that when some circumstances have been noted we can foresee that others will also be present. The progress of science consists in observing these interconnections and in showing with a patient ingenuity that the events of this evershifting world are but examples of a few general connections or relations called laws. To see what is general in what is particular and what is permanent in what is transitory is the aim of scientific thought. In the eye of science, the fall of an apple, the motion of a planet round a sun, and the clinging of the atmosphere to the earth are all seen as examples of the law of gravity. This possibility of disentangling the most complex evanescent circumstances into various examples of permanent laws is the controlling idea of modern thought.

Now let us think of the sort of laws which we want in order completely to realize this scientific ideal. Our knowledge of the particular facts of the world around us is gained from our sensations. We see, and hear, and taste, and smell, and feel hot and cold, and push, and rub, and ache, and tingle.

These are just our own personal sensations: my toothache cannot be your toothache, and my sight cannot be your sight. But we ascribe the origin of these sensations to relations between the things which form the external world. Thus the dentist extracts not the toothache but the tooth. And not only so, we also endeavour to imagine the world as one connected set of things which underlies all the perceptions of all people. There is not one world of things for my sensations and another for yours, but one world in which we both exist. It is the same tooth both for dentist and patient. Also we hear and we touch the same world as we see.

It is easy, therefore, to understand that we want to describe the connections between these external things in some way which does not depend on any particular sensations, nor even on all the sensations of any particular person. The laws satisfied by the course of events in the world of external things are to be described, if possible, in a neutral universal fashion, the same for blind men as for deaf men, and the same for beings with faculties beyond our ken as for normal human beings.

But when we have put aside our immediate sensations, the most serviceable part—from its clearness, definiteness, and universality—of what is left is composed of our general ideas of the abstract formal properties of things; in fact, the abstract mathematical ideas mentioned above. Thus it comes about that, step by step, and not realizing the full meaning of the process, mankind has been led to search for a mathematical description of the properties of the universe, because in this way only can a general idea of the course of events be formed, freed from reference to particular persons or to particular types of sensation. For example, it might be asked at dinner: "What was it which underlay my sensation of sight, yours of touch, and his of taste and smell?" the answer being "an apple." But in its final analysis, science seeks to describe an apple in terms of the positions and motions of molecules, a description which ignores me and you and him, and also ignores sight and touch and taste and smell. Thus mathematical ideas, because they are abstract, supply just what is wanted for a scientific description of the course of events.

This point has usually been misunderstood, from being thought of in too narrow a way. Pythagoras had a glimpse of it when he proclaimed that number was the source of all things. In modern times the belief that the ultimate explanation of all things was to be found in Newtonian mechanics was an adumbration of the truth that all science as it grows towards perfection becomes mathematical in its ideas.

CHAPTER II

VARIABLES

MATHEMATICS as a science commenced when first someone, probably a Greek, proved propositions about *any* things or about *some* things, without specification of definite particular things. These propositions were first enunciated by the Greeks for geometry; and, accordingly, geometry was the great Greek mathematical science. After the rise of geometry centuries passed away before algebra made a really effective start, despite some faint anticipations by the later Greek mathematicians.

The ideas of *any* and of *some* are introduced into algebra by the use of letters, instead of the definite numbers of arithmetic. Thus, instead of saying that $2 + 3 = 3 + 2$, in algebra we generalize and say that, if x and y stand for *any* two numbers, then $x + y = y + x$. Again, in the place of saying that $3 > 2$, we generalize and say that if x be *any* number there exists *some* number (or numbers) y such that $y > x$. We may remark in passing that this latter assumption—for when put in its strict ultimate form it is an assumption—is of vital importance, both to philosophy and to mathematics; for by it the notion of infinity is introduced. Perhaps it required the introduction of the arabic numerals, by which the use of letters as standing for definite numbers has been completely discarded in mathematics, in order to suggest to mathematicians the technical convenience of the use of letters for the ideas of *any* number and *some* number. The Romans would have stated the number of the year in which this is written in the form MDCCCCX., whereas we write it 1910, thus leaving the letters for the other usage. But this is merely a speculation. After the rise of algebra the differential calculus was invented by Newton and Leibniz, and then a pause in the progress of the philosophy of mathematical thought occurred so far as these notions are concerned; and it was not till within the last few years that it has been realized how fundamental *any* and *some* are to the very nature of mathematics, with the result of opening out still further subjects for mathematical exploration.

Let us now make some simple algebraic statements, with the object of understanding exactly how these fundamental ideas occur.

(1) For *any* number x, $x + 2 = 2 + x$;

(2) For *some* number x, $x + 2 = 3$;

(3) For *some* number x, $x + 2 > 3$.

The first point to notice is the possibilities contained in the meaning of *some*, as here used. Since $x+2 = 2+x$ for any number x, it is true for *some* number x. Thus, as here used, *any* implies *some* and *some* does not exclude *any*. Again, in the second example, there is, in fact, only one number x, such that $x + 2 = 3$, namely only the number 1. Thus the *some* may be one number only. But in the third example, any number x which is greater than 1 gives $x+2 > 3$. Hence there are an infinite number of numbers which answer to the *some* number in this case. Thus *some* may be anything between *any* and *one only*, including both these limiting cases.

It is natural to supersede the statements (2) and (3) by the questions:

(2′) For what number x is $x + 2 = 3$;

(3′) For what numbers x is $x + 2 > 3$.

Considering (2′), $x + 2 = 3$ is an equation, and it is easy to see that its solution is $x = 3 - 2 = 1$. When we have asked the question implied in the statement of the equation $x + 2 = 3$, x is called the unknown. The object of the solution of the equation is the determination of the unknown. Equations are of great importance in mathematics, and it seems as though (2′) exemplified a much more thoroughgoing and fundamental idea than the original statement (2). This, however, is a complete mistake. The idea of the undetermined "variable" as occurring in the use of "some" or "any" is the really important one in mathematics; that of the "unknown" in an equation, which is to be solved as quickly as possible, is only of subordinate use, though of course it is very important. One of the causes of the apparent triviality of much of elementary algebra is the preoccupation of the text-books with the solution of equations. The same remark applies to the solution of the inequality (3′) as compared to the original statement (3).

But the majority of interesting formulæ, especially when the idea of *some* is present, involve more than one variable. For example, the consideration of the pairs of numbers x and y (fractional or integral) which satisfy $x+y = 1$ involves the idea of two correlated variables, x and y. When two variables are present the same two main types of statement occur. For example, (1) for *any* pair of numbers, x and y, $x + y = y + x$, and (2) for *some* pairs of numbers, x and y, $x + y = 1$.

The second type of statement invites consideration of the aggregate of pairs of numbers which are bound together by some fixed relation—in the case given, by the relation $x+y = 1$. One use of formulæ of the first type, true for *any* pair of numbers, is that by them formulæ of the second type can be thrown into an indefinite number of equivalent forms. For example, the relation $x + y = 1$ is equivalent to the relations

$$y + x = 1, \quad (x - y) + 2y = 1, \quad 6x + 6y = 6,$$

and so on. Thus a skilful mathematician uses that equivalent form of the relation under consideration which is most convenient for his immediate purpose.

It is not in general true that, when a pair of terms satisfy some fixed relation, if one of the terms is given the other is also definitely determined. For example, when x and y satisfy $y^2 = x$, if $x = 4$, y can be ± 2, thus, for any positive value of x there are alternative values for y. Also in the relation $x + y > 1$, when either x or y is given, an indefinite number of values remain open for the other.

Again there is another important point to be noticed. If we restrict ourselves to positive numbers, integral or fractional, in considering the relation $x + y = 1$, then, if either x or y be greater than 1, there is no positive number which the other can assume so as to satisfy the relation. Thus the "field" of the relation for x is restricted to numbers less than 1, and similarly for the "field" open to y. Again, consider integral numbers only, positive or negative, and take the relation $y^2 = x$, satisfied by pairs of such numbers. Then whatever integral value is given to y, x can assume one corresponding integral value. So the "field" for y is unrestricted among these positive or negative integers. But the "field" for x is restricted in two ways. In the first place x must be positive, and in the second place, since y is to be integral, x must be a perfect square. Accordingly, the "field" of x is restricted to the set of integers 1^2, 2^2, 3^2, 4^2, and so on, *i.e.*, to 1, 4, 9, 16, and so on.

The study of the general properties of a relation between pairs of numbers is much facilitated by the use of a diagram constructed as follows:

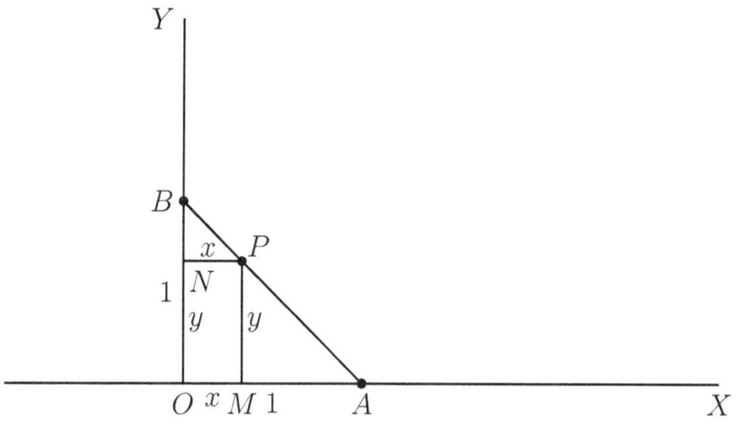

Fig. 1.

Draw two lines OX and OY at right angles; let any number x be represented by x units (in any scale) of length along OX, any number y by y units (in any scale) of length along OY. Thus if OM, along OX, be x units in length, and ON, along OY, be y units in length, by completing the parallelogram $OMPN$ we find a point P which corresponds to the pair of numbers x and y. To each point there corresponds one pair of numbers, and to each pair of numbers there corresponds one point. The pair of numbers are called the coordinates of the point. Then the points whose coordinates satisfy some fixed relation can be indicated in a convenient way, by drawing a line, if they all lie on a line, or by shading an area if they are all points in the area. If the relation can be represented by an equation such as $x+y=1$, or $y^2=x$, then the points lie on a line, which is straight in the former case and curved in the latter. For example, considering only positive numbers, the points whose coordinates satisfy $x+y=1$ lie on the straight line AB in Fig. 1, where $0A=1$ and $OB=1$. Thus this segment of the straight line AB gives a pictorial representation of the properties of the relation under the restriction to positive numbers.

Another example of a relation between two variables is afforded by considering the variations in the pressure and volume of a given mass of some gaseous substance—such as air or coal-gas or steam—at a constant temperature. Let v be the number of cubic feet in its volume and p its pressure in lb. weight per square inch. Then the law, known as Boyle's law, expressing the relation between p and v as both vary, is that the product pv is constant, always supposing that the temperature does not alter. Let us suppose, for example, that the quantity of the gas and its other circumstances are such that we can put $pv = 1$ (the exact number on the right-hand side of the equation makes no essential difference).

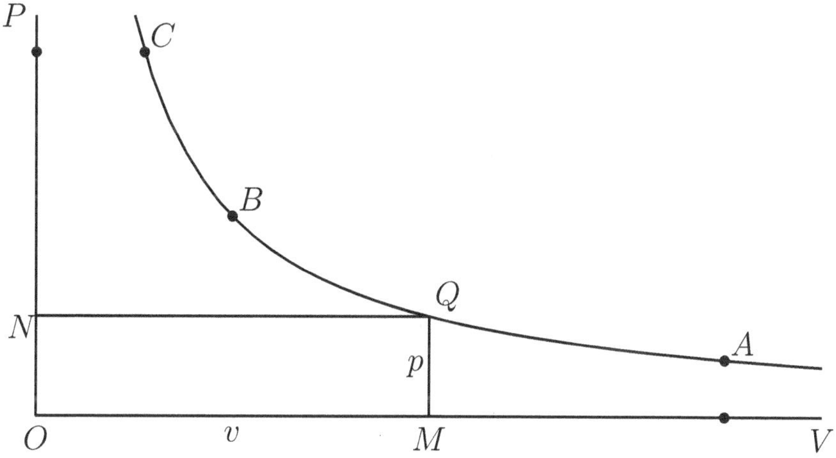

Fig. 2.

Then in Fig. 2 we take two lines, OV and OP, at right angles and draw OM along OV to represent v units of volume, and ON along OP to represent p units of pressure. Then the point Q, which is found by completing the parallelogram $OMQN$, represents the state of the gas when its volume is v cubic feet and its pressure is p lb. weight per square inch. If the circumstances of the portion of gas considered are such that $pv = 1$, then all these points Q which correspond to any possible state of this portion of gas must lie on the curved line ABC, which includes all points for which p and v are positive, and $pv = 1$. Thus this curved line gives a pictorial representation of the relation holding between the volume and the pressure. When the pressure is very big the corresponding point Q must be near C, or even beyond C on the undrawn part of the curve; then the volume will be very small. When the volume is big Q will be near to A, or beyond A; and then the pressure will be small. Notice that an engineer or a physicist may want to know the particular pressure corresponding to some definitely assigned volume. Then we have the case of determining the *unknown* p when v is a known number. But this is only in particular cases. In considering generally the properties of the gas and how it will behave, he has to have in his mind the general form of the whole curve ABC and its general properties. In other words the really fundamental idea is that of the pair of *variables* satisfying the relation $pv = 1$. This example illustrates how the idea of *variables* is fundamental, both in the applications as well as in the theory of mathematics.

CHAPTER III

METHODS OF APPLICATION

THE way in which the idea of variables satisfying a relation occurs in the applications of mathematics is worth thought, and by devoting some time to it we shall clear up our thoughts on the whole subject.

Let us start with the simplest of examples:—Suppose that building costs 1*s.* per cubic foot and that 20*s.* make £1. Then in all the complex circumstances which attend the building of a new house, amid all the various sensations and emotions of the owner, the architect, the builder, the workmen, and the onlookers as the house has grown to completion, this fixed correlation is by the law assumed to hold between the cubic content and the cost to the owner, namely that if x be the number of cubic feet, and £y the cost, then $20y = x$. This correlation of x and y is assumed to be true for the building of any house by any owner. Also, the volume of the house and the cost are not supposed to have been perceived or apprehended by any particular sensation or faculty, or by any particular man. They are stated in an abstract general way, with complete indifference to the owner's state of mind when he has to pay the bill.

Now think a bit further as to what all this means. The building of a house is a complicated set of circumstances. It is impossible to begin to apply the law, or to test it, unless amid the general course of events it is possible to recognize a definite set of occurrences as forming a particular instance of the building of a house. In short, we must know a house when we see it, and must recognize the events which belong to its building. Then amidst these events, thus isolated in idea from the rest of nature, the two elements of the cost and cubic content must be determinable; and when they are both determined, if the law be true, they satisfy the general formula

$$20y = x.$$

But is the law true? Anyone who has had much to do with building will know that we have here put the cost rather high. It is only for an expensive type of house that it will work out at this price. This brings out another point which must be made clear. While we are making mathematical calculations connected with the formula $20y = x$, it is indifferent to us whether the law be true or false. In fact, the very meanings assigned to x and y, as being a number of cubic feet and a number of pounds sterling, are indifferent. During the mathematical investigation we are, in fact, merely considering the properties of this correlation between a pair of variable numbers x and y. Our results will apply equally well, if we interpret y to mean a number of fishermen and x the number of fish caught, so that the assumed law is that on the average each fisherman catches twenty fish. The mathematical certainty of the investigation only attaches to the results considered as giving properties of the correlation $20y = x$ between the variable pair of numbers x and y. There is no mathematical certainty whatever about the cost of the actual building of any house. The law is not quite true and the result it gives will not be quite accurate. In fact, it may well be hopelessly wrong.

Now all this no doubt seems very obvious. But in truth with more complicated instances there is no more common error than to assume that, because prolonged and accurate mathematical calculations have been made, the application of the result to some fact of nature is absolutely certain. The conclusion of no argument can be more certain than the assumptions from which it starts. All mathematical calculations about the course of nature must start from some assumed law of nature, such, for instance, as the assumed law of the cost of building stated above. Accordingly, however accurately we have calculated that some event must occur, the doubt always remains—Is the law true? If the law states a precise result, almost certainly it is not precisely accurate; and thus even at the best the result, precisely as calculated, is not likely to occur. But then we have no faculty capable of observation with ideal precision, so, after all, our inaccurate laws may be good enough.

We will now turn to an actual case, that of Newton and the Law of Gravity. This law states that any two bodies attract one another with a force proportional to the product of their masses, and inversely proportional to the square of the distance between them. Thus if m and M are the masses of the two bodies, reckoned in lbs. say, and d miles is the distance between them, the force on either body, due to the attraction of the other and directed towards it, is proportional to $\frac{mM}{d^2}$; thus this force can be written as equal to $\frac{kmM}{d^2}$, where k is a definite number depending on the absolute magnitude of this attraction and also on the scale by which we choose to measure forces. It is easy to see that, if we wish to reckon in terms of forces such as the weight of a mass of 1 lb., the number which k represents must be extremely small; for when m and M and d are each put equal to 1, $\frac{kmM}{d^2}$ becomes the gravitational attraction of two equal masses of 1 lb. at the distance of one mile, and this is quite inappreciable.

However, we have now got our formula for the force of attraction. If we call this force F, it is $F = k\frac{mM}{d^2}$, giving the correlation between the variables F, m, M, and d. We all know the story of how it was found out. Newton, it states, was sitting in an orchard and watched the fall of an apple, and then the law of universal gravitation burst upon his mind. It may be that the final formulation of the law occurred to him in an orchard, as well as elsewhere—and he must have been somewhere. But for our purposes it is more instructive to dwell upon the vast amount of preparatory thought, the product of many minds and many centuries, which was necessary before this exact law could be formulated. In the first place, the mathematical habit of mind and the mathematical procedure explained in the previous two chapters had to be generated; otherwise Newton could never have thought of a formula representing the force between *any* two masses at *any* distance. Again, what are the meanings of the terms employed, Force, Mass, Distance? Take the easiest of these terms, Distance. It seems very obvious to us to conceive all material things as forming a definite geometrical whole, such that the distances of the various parts are measurable in terms of some unit length, such as a mile or a yard. This is almost the first aspect of a material structure which occurs to us. It is the gradual outcome of the study of geometry and of the theory of measurement. Even now, in certain cases, other modes of thought are convenient. In a mountainous country distances are often reckoned in hours. But leaving distance, the other terms, Force and Mass, are much more obscure. The exact comprehension of the ideas which Newton meant to convey by these words was of slow growth, and, indeed, Newton himself was the first man who had thoroughly mastered the true general principles of Dynamics.

Throughout the middle ages, under the influence of Aristotle, the science was entirely misconceived. Newton had the advantage of coming after a series of great men, notably Galileo, in Italy, who in the previous two centuries had reconstructed the science and had invented the right way of thinking about it. He completed their work. Then, finally, having the ideas of force, mass, and distance, clear and distinct in his mind, and realising their importance and their relevance to the fall of an apple and the motions of the planets, he hit upon the law of gravitation and proved it to be the formula always satisfied in these various motions.

The vital point in the application of mathematical formulæ is to have clear ideas and a correct estimate of their relevance to the phenomena under observation. No less than ourselves, our remote ancestors were impressed with the importance of natural phenomena and with the desirability of taking energetic measures to regulate the sequence of events. Under the influence of irrelevant ideas they executed elaborate religious ceremonies to aid the birth of the new moon, and performed sacrifices to save the sun during the crisis of an eclipse. There is no reason to believe that they were more stupid than we are. But at that epoch there had not been opportunity for the slow accumulation of clear and relevant ideas.

The sort of way in which physical sciences grow into a form capable of treatment by mathematical methods is illustrated by the history of the gradual growth of the science of electromagnetism. Thunderstorms are events on a grand scale, arousing terror in men and even animals. From the earliest times they must have been objects of wild and fantastic hypotheses, though it may be doubted whether our modern scientific discoveries in connection with electricity are not more astonishing than any of the magical explanations of savages. The Greeks knew that amber (Greek, electron) when rubbed would attract light and dry bodies. In 1600 A.D., Dr. Gilbert, of Colchester, published the first work on the subject in which any scientific method is followed.

He made a list of substances possessing properties similar to those of amber; he must also have the credit of connecting, however vaguely, electric and magnetic phenomena. At the end of the seventeenth and throughout the eighteenth century knowledge advanced. Electrical machines were made, sparks were obtained from them; and the Leyden Jar was invented, by which these effects could be intensified. Some organised knowledge was being obtained; but still no relevant mathematical ideas had been found out. Franklin, in the year 1752, sent a kite into the clouds and proved that thunderstorms were electrical.

Meanwhile from the earliest epoch (2634 B.C.) the Chinese had utilized the characteristic property of the compass needle, but do not seem to have connected it with any theoretical ideas. The really profound changes in human life all have their ultimate origin in knowledge pursued for its own sake. The use of the compass was not introduced into Europe till the end of the twelfth century A.D., more than 3000 years after its first use in China. The importance which the science of electromagnetism has since assumed in every department of human life is not due to the superior practical bias of Europeans, but to the fact that in the West electrical and magnetic phenomena were studied by men who were dominated by abstract theoretic interests.

The discovery of the electric current is due to two Italians, Galvani in 1780, and Volta in 1792. This great invention opened a new series of phenomena for investigation. The scientific world had now three separate, though allied, groups of occurrences on hand—the effects of "statical" electricity arising from frictional electrical machines, the magnetic phenomena, and the effects due to electric currents. From the end of the eighteenth century onwards, these three lines of investigation were quickly interconnected and the modern science of electromagnetism was constructed, which now threatens to transform human life.

Mathematical ideas now appear. During the decade 1780 to 1789, Coulomb, a Frenchman, proved that magnetic poles attract or repel each other, in proportion to the inverse square of their distances, and also that the same law holds for electric charges—laws curiously analogous to that of gravitation. In 1820, Öersted, a Dane, discovered that electric currents exert a force on magnets, and almost immediately afterwards the mathematical law of the force was correctly formulated by Ampère, a Frenchman, who also proved that two electric currents exerted forces on each other. "The experimental investigation by which Ampère established the law of the mechanical action between electric currents is one of the most brilliant achievements in science. The whole, theory and experiment, seems as if it had leaped, full-grown and full armed, from the brain of the 'Newton of Electricity.' It is perfect in form, and unassailable in accuracy, and it is summed up in a formula from which all the phenomena may be deduced, and which must always remain the cardinal formula of electro-dynamics."*

---

* *Electricity and Magnetism*, Clerk Maxwell, Vol. II., ch. iii.

The momentous laws of induction between currents and between currents and magnets were discovered by Michael Faraday in 1831–82. Faraday was asked: "What is the use of this discovery?" He answered: "What is the use of a child—it grows to be a man." Faraday's child has grown to be a man and is now the basis of all the modern applications of electricity. Faraday also reorganized the whole theoretical conception of the science. His ideas, which had not been fully understood by the scientific world, were extended and put into a directly mathematical form by Clerk Maxwell in 1873. As a result of his mathematical investigations, Maxwell recognized that, under certain conditions, electrical vibrations ought to be propagated. He at once suggested that the vibrations which form light are electrical. This suggestion has since been verified, so that now the whole theory of light is nothing but a branch of the great science of electricity. Also Herz, a German, in 1888, following on Maxwell's ideas, succeeded in producing electric vibrations by direct electrical methods. His experiments are the basis of our wireless telegraphy.

In more recent years even more fundamental discoveries have been made, and the science continues to grow in theoretic importance and in practical interest. This rapid sketch of its progress illustrates how, by the gradual introduction of the relevant theoretic ideas, suggested by experiment and themselves suggesting fresh experiments, a whole mass of isolated and even trivial phenomena are welded together into one coherent science, in which the results of abstract mathematical deductions, starting from a few simple assumed laws, supply the explanation to the complex tangle of the course of events.

Finally, passing beyond the particular sciences of electromagnetism and light, we can generalize our point of view still further, and direct our attention to the growth of mathematical physics considered as one great chapter of scientific thought. In the first place, what in the barest outlines is the story of its growth?

It did not begin as one science, or as the product of one band of men. The Chaldean shepherds watched the skies, the agents of Government in Mesopotamia and Egypt measured the land, priests and philosophers brooded on the general nature of all things. The vast mass of the operations of nature appeared due to mysterious unfathomable forces. "The wind bloweth where it listeth" expresses accurately the blank ignorance then existing of any stable rules followed in detail by the succession of phenomena. In broad outline, then as now, a regularity of events was patent. But no minute tracing of their interconnection was possible, and there was no knowledge how even to set about to construct such a science.

Detached speculations, a few happy or unhappy shots at the nature of things, formed the utmost which could be produced.

Meanwhile land-surveys had produced geometry, and the observations of the heavens disclosed the exact regularity of the solar system. Some of the later Greeks, such as Archimedes, had just views on the elementary phenomena of hydrostatics and optics. Indeed, Archimedes, who combined a genius for mathematics with a physical insight, must rank with Newton, who lived nearly two thousand years later, as one of the founders of mathematical physics. He lived at Syracuse, the great Greek city of Sicily. When the Romans besieged the town (in 212 to 210 B.C.), he is said to have burned their ships by concentrating on them, by means of mirrors, the sun's rays. The story is highly improbable, but is good evidence of the reputation which he had gained among his contemporaries for his knowledge of optics. At the end of this siege he was killed. According to one account given by Plutarch, in his life of Marcellus, he was found by a Roman soldier absorbed in the study of a geometrical diagram which he had traced on the sandy floor of his room. He did not immediately obey the orders of his captor, and so was killed. For the credit of the Roman generals it must be said that the soldiers had orders to spare him. The internal evidence for the other famous story of him is very strong; for the discovery attributed to him is one eminently worthy of his genius for mathematical and physical research. Luckily, it is simple enough to be explained here in detail. It is one of the best easy examples of the method of application of mathematical ideas to physics.

Hicro, King of Syracuse, had sent a quantity of gold to some goldsmith to form the material of a crown. He suspected that the craftsmen had abstracted some of the gold and had supplied its place by alloying the remainder with some baser metal. Hiero sent the crown to Archimedes and asked him to test it. In these days an indefinite number of chemical tests would be available. But then Archimedes had to think out the matter afresh. The solution flashed upon him as he lay in his bath. He jumped up and ran through the streets to the palace, shouting *Eureka! Eureka!* (I have found it, I have found it). This day, if we knew which it was, ought to be celebrated as the birthday of mathematical physics; the science came of age when Newton sat in his orchard. Archimedes had in truth made a great discovery. He saw that a body when immersed in water is pressed upwards by the surrounding water with a resultant force equal to the weight of the water it displaces. This law can be proved theoretically from the mathematical principles of hydrostatics and can also be verified experimentally. Hence, if W lb. be the weight of the crown, as weighed in air, and w lb. be the weight of the water which it displaces when completely immersed, $W - w$ would be the extra upward force necessary to sustain the crown as it hung in water.

Now, this upward force can easily be ascertained by weighing the body as it hangs in water, as shown in the annexed figure. If the weights in the right-hand scale come to F lb., then the apparent weight of the crown in water is F lb.; and we thus have

$$F = W - w$$

and thus

$$w = W - F,$$

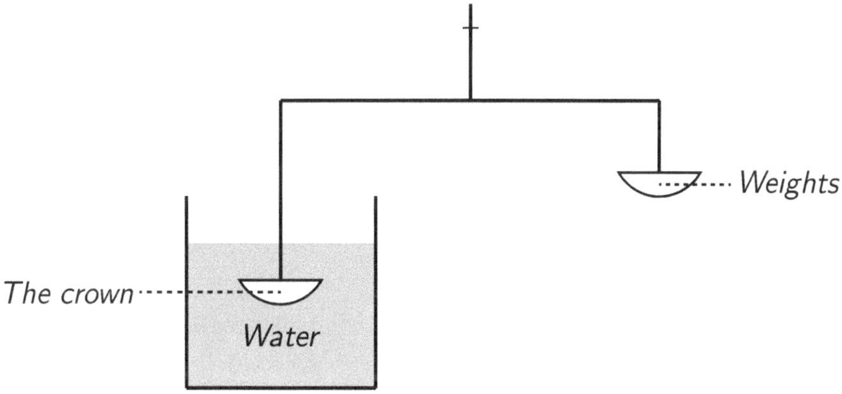

Fig. 3.

and

$$(A) \qquad \frac{W}{w} = \frac{W}{W - F}$$

where W and F are determined by the easy, and fairly precise, operation of weighing. Hence, by equation (A), $\frac{W}{w}$ is known. But $\frac{W}{w}$ is the ratio of the weight of the crown to the weight of an equal volume of water. This ratio is the same for any lump of metal of the same material: it is now called the specific gravity of the material, and depends only on the intrinsic nature of the substance and not on its shape or quantity. Thus to test if the crown were of gold, Archimedes had only to take a lump of indisputably pure gold and find its specific gravity by the same process. If the two specific gravities agreed, the crown was pure; if they disagreed, it was debased.

This argument has been given at length, because not only is it the first precise example of the application of mathematical ideas to physics, but also because it is a perfect and simple example of what must be the method and spirit of the science for all time.

The death of Archimedes by the hands of a Roman soldier is symbolical of a world-change of the first magnitude: the theoretical Greeks, with their love of abstract science, were superseded in the leadership of the European world by the practical Romans. Lord Beaconsfield, in one of his novels, has defined a practical man as a man who practises the errors of his forefathers. The Romans were a great race, but they were cursed with the sterility which waits upon practicality. They did not improve upon the knowledge of their forefathers, and all their advances were confined to the minor technical details of engineering. They were not dreamers enough to arrive at new points of view, which could give a more fundamental control over the forces of nature. No Roman lost his life because he was absorbed in the contemplation of a mathematical diagram.

CHAPTER IV

DYNAMICS

THE world had to wait for eighteen hundred years till the Greek mathematical physicists found successors. In the sixteenth and seventeenth centuries of our era great Italians, in particular Leonardo da Vinci, the artist (born 1452, died 1519), and Galileo (born 1564, died 1642), rediscovered the secret, known to Archimedes, of relating abstract mathematical ideas with the experimental investigation of natural phenomena. Meanwhile the slow advance of mathematics and the accumulation of accurate astronomical knowledge had placed natural philosophers in a much more advantageous position for research. Also the very egoistic self-assertion of that age, its greediness for personal experience, led its thinkers to want to see for themselves what happened; and the secret of the relation of mathematical theory and experiment in inductive reasoning was practically discovered. It was an act eminently characteristic of the age that Galileo, a philosopher, should have dropped the weights from the leaning tower of Pisa. There are always men of thought and men of action; mathematical physics is the product of an age which combined in the same men impulses to thought with impulses to action.

This matter of the dropping of weights from the tower marks picturesquely an essential step in knowledge, no less a step than the first attainment of correct ideas on the science of dynamics, the basal science of the whole subject. The particular point in dispute was as to whether bodies of different weights would fall from the same height in the same time. According to a dictum of Aristotle, universally followed up to that epoch, the heavier weight would fall the quicker. Galileo affirmed that they would fall in the same time, and proved his point by dropping weights from the top of the leaning tower. The apparent exceptions to the rule all arise when, for some reason, such as extreme lightness or great speed, the air resistance is important. But neglecting the air the law is exact.

Galileo's successful experiment was not the result of a mere lucky guess. It arose from his correct ideas in connection with inertia and mass. The first law of motion, as following Newton we now enunciate it, is—Every body continues in its state of rest or of uniform motion in a straight line, except so far as it is compelled by impressed force to change that state. This law is more than a dry formula: it is also a pæan of triumph over defeated heretics. The point at issue can be understood by deleting from the law the phrase "or of uniform motion in a straight line." We there obtain what might be taken as the Aristotelian opposition formula: "Every body continues in its state of rest except so far as it is compelled by impressed force to change that state."

In this last false formula it is asserted that, apart from force, a body continues in a state of rest; and accordingly that, if a body is moving, a force is required to sustain the motion; so that when the force ceases, the motion ceases.

The true Newtonian law takes diametrically the opposite point of view. The state of a body unacted on by force is that of uniform motion in a straight line, and no external force or influence is to be looked for as the cause, or, if you like to put it so, as the invariable accompaniment of this uniform rectilinear motion. Rest is merely a particular case of such motion, merely when the velocity is and remains zero. Thus, when a body is moving, we do not seek for any external influence except to explain changes in the rate of the velocity or changes in its direction. So long as the body is moving at the same rate and in the same direction there is no need to invoke the aid of any forces.

The difference between the two points of view is well seen by reference to the theory of the motion of the planets. Copernicus, a Pole, born at Thorn in West Prussia (born 1473, died 1543), showed how much simpler it was to conceive the planets, including the earth as revolving round the sun in orbits which are nearly circular; and later, Kepler, a German mathematician, in the year 1609 proved that, in fact, the orbits are practically ellipses, that is, a special sort of oval curves which we will consider later in more detail. Immediately the question arose as to what are the forces which preserve the planets in this motion. According to the old false view, held by Kepler, the actual velocity itself required preservation by force. Thus he looked for tangential forces as in the accompanying figure (4). But according to the Newtonian law, apart from some force the planet would move for ever with its existing velocity in a straight line, and

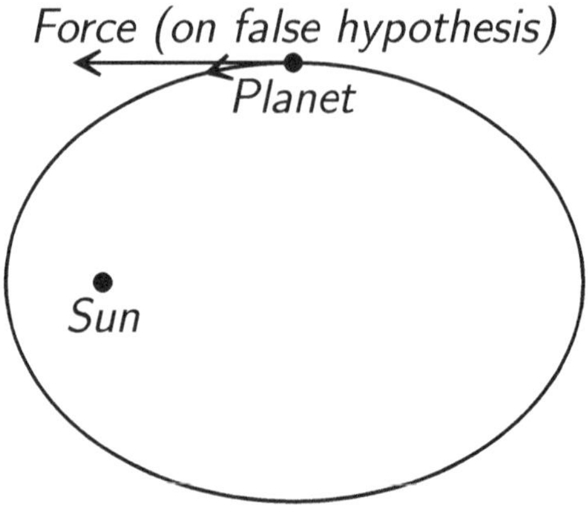

Fig. 4.

thus depart entirely from the sun. Newton, therefore, had to search for a force which would bend the motion round into

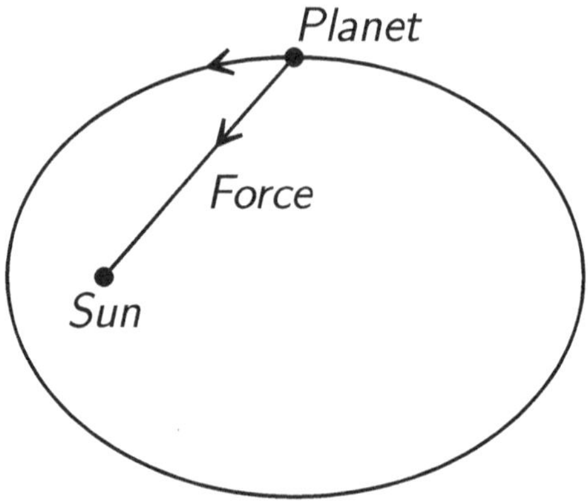

Fig. 5.

its elliptical orbit. This he showed must be a force directed towards the sun as in the next figure (5). In fact, the force is the gravitational attraction of the sun acting according to the law of the inverse square of the distance, which has been stated above.

The science of mechanics rose among the Greeks from a consideration of the theory of the mechanical advantage obtained by the use of a lever, and also from a consideration of various problems connected with the weights of bodies. It was finally put on its true basis at the end of the sixteenth and during the seventeenth centuries, as the preceding account shows, partly with the view of explaining the theory of falling bodies, but chiefly in order to give a scientific theory of planetary motions. But since those days dynamics has taken upon itself a more ambitious task, and now claims to be the ultimate science of which the others are but branches. The claim amounts to this: namely, that the various qualities of things perceptible to the senses are merely our peculiar mode of appreciating changes in position on the part of things existing in space. For example, suppose we look at Westminster Abbey. It has been standing there, grey and immovable, for centuries past. But, according to modern scientific theory, that greyness, which so heightens our sense of the immobility of the building, is itself nothing but our way of appreciating the rapid motions of the ultimate molecules, which form the outer surface of the building and communicate vibrations to a substance called the ether. Again we lay our hands on its stones and note their cool, even temperature, so symbolic of the quiet repose of the building. But this feeling of temperature simply marks our sense of the transfer of heat from the hand to the stone, or from the stone to the hand; and, according to modern science, heat is nothing but the agitation of the molecules of a body. Finally, the organ begins playing, and again sound is nothing but the result of motions of the air striking on the drum of the ear.

Thus the endeavour to give a dynamical explanation of phenomena is the attempt to explain them by statements of the general form, that such and such a substance or body was in this place and is now in that place. Thus we arrive at the great basal idea of modern science, that all our sensations are the result of comparisons of the changed configurations of things in space at various times. It follows therefore, that the laws of motion, that is, the laws of the changes of configurations of things, are the ultimate laws of physical science.

In the application of mathematics to the investigation of natural philosophy, science does systematically what ordinary thought does casually. When we talk of a chair, we usually mean something which we have been seeing or feeling in some way; though most of our language will presuppose that there is something which exists independently of our sight or feeling. Now in mathematical physics the opposite course is taken. The chair is conceived without any reference to anyone in particular, or to any special modes of perception. The result is that the chair becomes in thought a set of molecules in space, or a group of electrons, a portion of the ether in motion, or however the current scientific ideas describe it. But the point is that science reduces the chair to things moving in space and influencing each other's motions. Then the various elements or factors which enter into a set of circumstances, as thus conceived, are merely the things, like lengths of lines, sizes of angles, areas, and volumes, by which the positions of bodies in space can be settled. Of course, in addition to these geometrical elements the fact of motion and change necessitates the introduction of the rates of changes of such elements, that is to say, velocities, angular velocities, accelerations, and suchlike things. Accordingly, mathematical physics deals with correlations between variable numbers which are supposed to represent the correlations which exist in nature between the measures of these geometrical elements and of their rates of change. But always the mathematical laws deal with variables, and it is only in the occasional testing of the laws by reference to experiments, or in the use of the laws for special predictions that definite numbers are substituted.

The interesting point about the world as thus conceived in this abstract way throughout the study of mathematical physics, where only the positions and shapes of things are considered together with their changes, is that the events of such an abstract world are sufficient to "explain" our sensations. When we hear a sound, the molecules of the air have been agitated in a certain way: given the agitation, or air-waves as they are called, all normal people hear sound; and if there are no air-waves, there is no sound. And, similarly, a physical cause or origin, or parallel event (according as different people might like to phrase it) underlies our other sensations. Our very thoughts appear to correspond to conformations and motions of the brain; injure the brain and you injure the thoughts. Meanwhile the events of this physical universe succeed each other according to the mathematical laws which ignore all special sensations and thoughts and emotions.

Now, undoubtedly, this is the general aspect of the relation of the world of mathematical physics to our emotions, sensations, and thoughts; and a great deal of controversy has been occasioned by it and much ink spilled. We need only make one remark. The whole situation has arisen, as we have seen, from the endeavour to describe an external world "explanatory" of our various individual sensations and emotions, but a world also, not essentially dependent upon any particular sensations or upon any particular individual. Is such a world merely but one huge fairy tale? But fairy tales are fantastic and arbitrary: if in truth there be such a world, it ought to submit itself to an exact description, which determines accurately its various parts and their mutual relations. Now, to a large degree, this scientific world does submit itself to this test and allow its events to be explored and predicted by the apparatus of abstract mathematical ideas. It certainly seems that here we have an inductive verification of our initial assumption. It must be admitted that no inductive proof is conclusive; but if the whole idea of a world which has existence independently of our particular perceptions of it be erroneous, it requires careful explanation why the attempt to characterise it, in terms of that mathematical remnant of our ideas which would apply to it, should issue in such a remarkable success.

It would take us too far afield to enter into a detailed explanation of the other laws of motion. The remainder of this chapter must be devoted to the explanation of remarkable ideas which are fundamental, both to mathematical physics and to pure mathematics: these are the ideas of vector quantities and the parallelogram law for vector addition. We have seen that the essence of motion is that a body was at A and is now at C. This transference from A to C requires two distinct elements to be settled before it is completely determined, namely its magnitude (*i.e.* the length AC) and its direction. Now anything, like this transference, which is completely given by the determination of a magnitude and

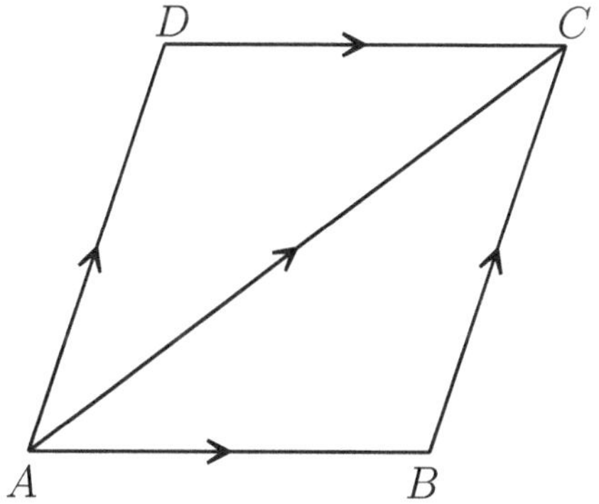

Fig. 6.

a direction is called a vector. For example, a velocity requires for its definition the assignment of a magnitude and of a direction. It must be of so many miles per hour in such and such a direction. The existence and the independence of these two elements in the determination of a velocity are well illustrated by the action of the captain of a ship, who communicates with different subordinates respecting them: he tells the chief engineer the number of knots at which he is to steam, and the helmsman the compass bearing of the course which he is to keep. Again the rate of change of velocity, that is velocity added per unit time, is also a vector quantity: it is called the acceleration. Similarly a force in the dynamical sense is another vector quantity. Indeed, the vector nature of forces follows at once according to dynamical principles from that of velocities and accelerations; but this is a point which we need not go into. It is sufficient here to say that a force acts on a body with a certain magnitude in a certain direction.

Now all vectors can be graphically represented by straight lines. All that has to be done is to arrange: (i) a scale according to which units of length correspond to units of magnitude of the vector—for example, one inch to a velocity of 10 miles per hour in the case of velocities, and one inch to a force of 10 tons weight in the case of forces—and (ii) a direction of the line on the diagram corresponding to the direction of the vector. Then a line drawn with the proper number of inches of length in the proper direction represents the required vector on the arbitrarily assigned scale of magnitude. This diagrammatic representation of vectors is of the first importance. By its aid we can enunciate the famous "parallelogram law" for the addition of vectors of the same kind but in different directions.

Consider the vector AC in figure 6 as representative of the changed position of a body from A to C: we will call this the vector of transportation. It will be noted that, if the reduction of physical phenomena to mere changes in positions, as explained above, is correct, all other types of physical vectors are really reducible in some way or other to this single type. Now the final transportation from A to C is equally well effected by a transportation from A to B and a transportation from B to C, or, completing the parallelogram $ABCD$, by a transportation from A to D and a transportation from D to C. These transportations as thus successively applied are said to be added together. This is simply a definition of what we mean by the addition of transportations. Note further that, considering parallel lines as being lines drawn in the same direction, the transportations B to C and A to D may be conceived as the same transportation applied to bodies in the two initial positions B and A. With this conception we may talk of the transportation A to D as applied to a body in any position, for example at B. Thus we may say that the transportation A to C can be conceived as the sum of the two transportations A to B and A to D applied in any order. Here we have the parallelogram law for the addition of transportations: namely, if the transportations are A to B and A to D, complete the parallelogram $ABCD$, and then the sum of the two is the diagonal AC.

All this at first sight may seem to be very artificial. But

it must be observed that nature itself presents us with the idea. For example, a steamer is moving in the direction AD (*cf.* fig. 6) and a man walks across its deck. If the steamer were still, in one minute he would arrive at B; but during that minute his starting point A on the deck has moved to D, and his path on the deck has moved from AB to DC. So that, in fact, his transportation has been from A to C over the surface of the sea. It is, however, presented to us analysed into the sum of two transportations, namely, one from A to B relatively to the steamer, and one from A to D which is the transportation of the steamer.

By taking into account the element of time, namely one minute, this diagram of the man's transportation AC represents his velocity. For if AC represented so many feet of transportation, it now represents a transportation of so many feet per minute, that is to say, it represents the velocity of the man. Then AB and AD represent two velocities, namely, his velocity relatively to the steamer, and the velocity of the steamer, whose "sum" makes up his complete velocity. It is evident that diagrams and definitions concerning transportations are turned into diagrams and definitions concerning velocities by conceiving the diagrams as representing transportations per unit time. Again, diagrams and definitions concerning velocities are turned into diagrams and definitions concerning accelerations by conceiving the diagrams as representing velocities added per unit time.

Thus by the addition of vector velocities and of vector accelerations, we mean the addition according to the paral-

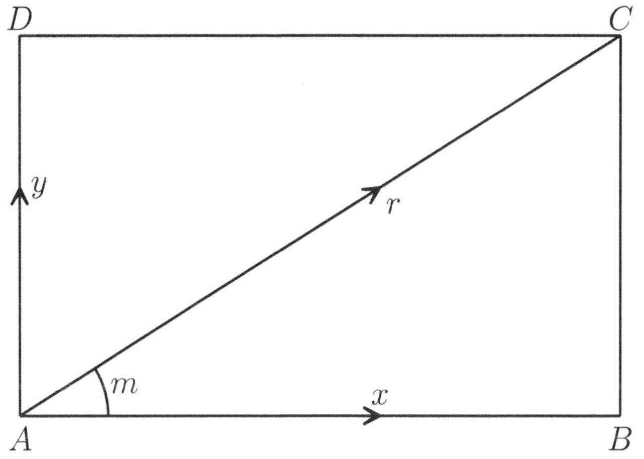

Fig. 7.

lelogram law.

Also, according to the laws of motion a force is fully represented by the vector acceleration it produces in a body of given mass. Accordingly, forces will be said to be added when their joint effect is to be reckoned according to the parallelogram law.

Hence for the fundamental vectors of science, namely transportations, velocities, and forces, the addition of any two of the same kind is the production of a "resultant" vector according to the rule of the parallelogram law.

By far the simplest type of parallelogram is a rectangle, and in pure mathematics it is the relation of the single vector AC to the two component vectors, AB and AD, at right angles (*cf.* fig. 7), which is continually recurring. Let

x, y, and r units represent the lengths of AB, AD, and AC, and let m units of angle represent the magnitude of the angle BAC. Then the relations between x, y, r, and m, in all their many aspects are the continually recurring topic of pure mathematics; and the results are of the type required for application to the fundamental vectors of mathematical physics. This diagram is the chief bridge over which the results of pure mathematics pass in order to obtain application to the facts of nature.

CHAPTER V

THE SYMBOLISM OF MATHEMATICS

WE now return to pure mathematics, and consider more closely the apparatus of ideas out of which the science is built. Our first concern is with the symbolism of the science, and we start with the simplest and universally known symbols, namely those of arithmetic.

Let us assume for the present that we have sufficiently clear ideas about the integral numbers, represented in the Arabic notation by 0, 1, 2, ..., 9, 10, 11, ..., 100, 101, ... and so on. This notation was introduced into Europe through the Arabs, but they apparently obtained it from Hindoo sources. The first known work* in which it is systematically explained is a work by an Indian mathematician, Bhaskara (born 1114 A.D.). But the actual numerals can be traced back to the seventh century of our era, and perhaps were originally invented in Tibet. For our present purposes, however, the history of the notation is a detail. The interesting point to notice is the admirable illustration which this numeral system affords of the enormous importance of a good notation. By relieving the brain of all unnecessary work, a good notation sets it free to concentrate on more advanced problems, and in effect increases the mental

*For the detailed historical facts relating to pure mathematics, I am chiefly indebted to *A Short History of Mathematics*, by W. W. R. Ball.

power of the race. Before the introduction of the Arabic notation, multiplication was difficult, and the division even of integers called into play the highest mathematical faculties. Probably nothing in the modern world would have more astonished a Greek mathematician than to learn that, under the influence of compulsory education, a large proportion of the population of Western Europe could perform the operation of division for the largest numbers. This fact would have seemed to him a sheer impossibility. The consequential extension of the notation to decimal fractions was not accomplished till the seventeenth century. Our modern power of easy reckoning with decimal fractions is the almost miraculous result of the gradual discovery of a perfect notation.

Mathematics is often considered a difficult and mysterious science, because of the numerous symbols which it employs. Of course, nothing is more incomprehensible than a symbolism which we do not understand. Also a symbolism, which we only partially understand and are unaccustomed to use, is difficult to follow. In exactly the same way the technical terms of any profession or trade are incomprehensible to those who have never been trained to use them. But this is not because they are difficult in themselves. On the contrary they have invariably been introduced to make things easy. So in mathematics, granted that we are giving any serious attention to mathematical ideas, the symbolism is invariably an immense simplification. It is not only of practical use, but is of great interest. For it represents an analysis of the ideas of the subject and an almost pictorial representation of their relations to each other. If anyone doubts the utility of symbols, let him write out in full, without any symbol whatever, the whole meaning of the following equations which represent some of the fundamental laws of algebra*:—

$$(1) \qquad x + y = y + x,$$

$$(2) \qquad (x + y) + z = x + (y + z),$$

$$(3) \qquad x \times y = y \times x,$$

$$(4) \qquad (x \times y) \times z = x \times (y \times z),$$

$$(5) \qquad x \times (y + z) = (x \times y) + (x \times z).$$

Here (1) and (2) are called the commutative and associative laws for addition, (3) and (4) are the commutative and associative laws for multiplication, and (5) is the distributive law relating addition and multiplication. For example, without symbols, (1) becomes: If a second number be added to any given number the result is the same as if the first given number had been added to the second number.

This example shows that, by the aid of symbolism, we can make transitions in reasoning almost mechanically by the eye, which otherwise would call into play the higher faculties of the brain.

It is a profoundly erroneous truism, repeated by all copy-books and by eminent people when they are making speeches, that we should cultivate the habit of thinking of

---

* *Cf.* Note A, p. 207.

what we are doing. The precise opposite is the case. Civilization advances by extending the number of important operations which we can perform without thinking about them. Operations of thought are like cavalry charges in a battle—they are strictly limited in number, they require fresh horses, and must only be made at decisive moments.

One very important property for symbolism to possess is that it should be concise, so as to be visible at one glance of the eye and to be rapidly written. Now we cannot place symbols more concisely together than by placing them in immediate juxtaposition. In a good symbolism therefore, the juxtaposition of important symbols should have an important meaning. This is one of the merits of the Arabic notation for numbers; by means of ten symbols, 0, 1, 2, 3, 4, 5, 6, 7, 8, 9, and by simple juxtaposition it symbolizes any number whatever. Again in algebra, when we have two variable numbers x and y, we have to make a choice as to what shall be denoted by their juxtaposition xy. Now the two most important ideas on hand are those of addition and multiplication. Mathematicians have chosen to make their symbolism more concise by defining xy to stand for $x \times y$. Thus the laws (3), (4), and (5) above are in general written,

$$xy = yx, \quad (xy)z = x(yz), \quad x(y + z) = xy + xz,$$

thus securing a great gain in conciseness. The same rule of symbolism is applied to the juxtaposition of a definite number and a variable: we write $3x$ for $3 \times x$, and $30x$ for $30 \times x$.

It is evident that in substituting definite numbers for the variables some care must be taken to restore the $\times$, so as not to conflict with the Arabic notation. Thus when we substitute 2 for x and 3 for y in xy, we must write 2×3 for xy, and not 23 which means $20 + 3$.

It is interesting to note how important for the development of science a modest-looking symbol may be. It may stand for the emphatic presentation of an idea, often a very subtle idea, and by its existence make it easy to exhibit the relation of this idea to all the complex trains of ideas in which it occurs. For example, take the most modest of all symbols, namely, 0, which stands for the *number* zero. The Roman notation for numbers had no symbol for zero, and probably most mathematicians of the ancient world would have been horribly puzzled by the idea of the number zero. For, after all, it is a very subtle idea, not at all obvious. A great deal of discussion on the meaning of the zero of quantity will be found in philosophic works. Zero is not, in real truth, more difficult or subtle in idea than the other cardinal numbers. What do we mean by 1 or by 2, or by 3? But we are familiar with the use of these ideas, though we should most of us be puzzled to give a clear analysis of the simpler ideas which go to form them. The point about zero is that we do not need to use it in the operations of daily life. No one goes out to buy zero fish. It is in a way the most civilized of all the cardinals, and its use is only forced on us by the needs of cultivated modes of thought. Many important services are rendered by the symbol 0, which stands for the number zero.

The symbol developed in connection with the Arabic notation for numbers of which it is an essential part. For in that notation the value of a digit depends on the position in which it occurs. Consider, for example, the digit 5, as occurring in the numbers 25, 51, 3512, 5213. In the first number 5 stands for five, in the second number 5 stands for fifty, in the third number for five hundred, and in the fourth number for five thousand. Now, when we write the number fifty-one in the symbolic form 51, the digit 1 pushes the digit 5 along to the second place (reckoning from right to left) and thus gives it the value fifty. But when we want to symbolize fifty by itself, we can have no digit 1 to perform this service; we want a digit in the units place to add nothing to the total and yet to push the 5 along to the second place. This service is performed by 0, the symbol for zero. It is extremely probable that the men who introduced for this purpose had no definite conception in their minds of the number zero. They simply wanted a mark to symbolize the fact that nothing was contributed by the digit's place in which it occurs. The idea of zero probably took shape gradually from a desire to assimilate the meaning of this mark to that of the marks, 1, 2, ..., 9, which do represent cardinal numbers. This would not represent the only case in which a subtle idea has been introduced into mathematics by a symbolism which in its origin was dictated by practical convenience.

Thus the first use of 0 was to make the arable notation possible—no slight service. We can imagine that when it had been introduced for this purpose, practical men, of the sort who dislike fanciful ideas, deprecated the silly habit of identifying it with a number zero. But they were wrong, as such men always are when they desert their proper function of masticating food which others have prepared. For the next service performed by the symbol 0 essentially depends upon assigning to it the function of representing the number zero.

This second symbolic use is at first sight so absurdly simple that it is difficult to make a beginner realize its importance. Let us start with a simple example. In Chapter II. we mentioned the correlation between two variable numbers x and y represented by the equation $x + y = 1$. This can be represented in an indefinite number of ways; for example, $x = 1 - y$, $y = 1 - x$, $2x + 3y - 1 = x + 2y$, and so on. But the important way of stating it is

$$x + y - 1 = 0.$$

Similarly the important way of writing the equation $x = 1$ is $x - 1 = 0$, and of representing the equation $3x - 2 = 2x^2$ is $2x^2 - 3x + 2 = 0$. The point is that all the symbols which represent variables, *e.g.* x and y, and the symbols representing some definite number other than zero, such as 1 or 2 in the examples above, are written on the left-hand side, so that the whole left-hand side is equated to the number zero. The first man to do this is said to have been Thomas Harriot, born at Oxford in 1560 and died in 1621. But what is the importance of this simple symbolic procedure? It made possible the growth of the modern conception of *algebraic form.*

This is an idea to which we shall have continually to recur; it is not going too far to say that no part of modern mathematics can be properly understood without constant recurrence to it. The conception of form is so general that it is difficult to characterize it in abstract terms. At this stage we shall do better merely to consider examples. Thus the equations $2x-3=0$, $x-1=0$, $5x-6=0$, are all equations of the same form, namely, equations involving one unknown x, which is not multiplied by itself, so that x^2, x^3, etc., do not appear. Again $3x^2 - 2x + 1 = 0$, $x^2 - 3x + 2 = 0$, $x^2 - 4 = 0$, are all equations of the same form, namely, equations involving one unknown x in which $x \times x$, that is x^2, appears. These equations are called quadratic equations. Similarly cubic equations, in which x^3 appears, yield another form, and so on. Among the three quadratic equations given above there is a minor difference between the last equation, $x^2 - 4 = 0$, and the preceding two equations, due to the fact that x (as distinct from x^2) does not appear in the last and does in the other two. This distinction is very unimportant in comparison with the great fact that they are all three quadratic equations.

Then further there are the forms of equation stating correlations between two variables; for example, $x + y - 1 = 0$, $2x + 3y - 8 = 0$, and so on. These are examples of what is called the *linear* form of equation. The reason for this name of "linear" is that the graphic method of representation, which is explained at the end of Chapter II., always represents such equations by a straight line. Then there are other forms for two variables—for example, the quadratic form, the cubic form, and so on. But the point which we here insist upon is that this study of form is facilitated, and, indeed, made possible, by the standard method of writing equations with the symbol 0 on the right-hand side.

There is yet another function performed by 0 in relation to the study of form. Whatever number x may be, $0 \times x = 0$, and $x+0 = x$. By means of these properties minor differences of form can be assimilated. Thus the difference mentioned above between the quadratic equations $x^2 - 3x + 2 = 0$, and $x^2 - 4 = 0$, can be obliterated by writing the latter equation in the form $x^2 + (0 \times x) - 4 = 0$. For, by the laws stated above, $x^2 + (0 \times x) - 4 = x^2 + 0 - 4 = x^2 - 4$. Hence the equation $x^2 - 4 = 0$ is merely representative of a particular class of quadratic equations and belongs to the same general form as does $x^2 - 3x + 2 = 0$.

For these three reasons the symbol 0, representing the number zero, is essential to modern mathematics. It has rendered possible types of investigation which would have been impossible without it.

The symbolism of mathematics is in truth the outcome of the general ideas which dominate the science. We have now two such general ideas before us, that of the variable and that of algebraic form. The junction of these concepts has imposed on mathematics another type of symbolism almost quaint in its character, but none the less effective. We have seen that an equation involving two variables, x and y, represents a particular correlation between the pair of variables. Thus $x+y-1=0$ represents one definite correlation, and $3x+2y-5=0$ represents another definite correlation between the variables x and y; and both correlations have the form of what we have called linear correlations. But now, how can we represent *any* linear correlation between the variable numbers x and y? Here we want to symbolize *any* linear correlation; just as x symbolizes *any* number. This is done by turning the numbers which occur in the definite correlation $3x+2y-5=0$ into letters. We obtain $ax+by-c=0$. Here a, b, c, stand for variable numbers just as do x and y: but there is a difference in the use of the two sets of variables. We study the general properties of the relationship between x and y while a, b, and c have unchanged values. We do not determine what the values of a, b, and c are; but whatever they are, they remain fixed while we study the relation between the variables x and y for the whole group of possible values of x and y. But when we have obtained the properties of this correlation, we note that, because a, b, and c have not in fact been determined, we have proved properties which must belong to *any* such relation. Thus, by now varying a, b, and c, we arrive at the idea that $ax+by-c=0$ represents a variable linear correlation between x and y. In comparison with x and y, the three variables a, b, and c are called constants. Variables used in this way are sometimes also called parameters.

Now, mathematicians habitually save the trouble of explaining which of their variables are to be treated as "constants," and which as variables, considered as correlated in their equations, by using letters at the end of the alphabet for the "variable" variables, and letters at the beginning of the alphabet for the "constant" variables, or parameters. The two systems meet naturally about the middle of the alphabet. Sometimes a word or two of explanation is necessary; but as a matter of fact custom and common sense are usually sufficient, and surprisingly little confusion is caused by a procedure which seems so lax.

The result of this continual elimination of definite numbers by successive layers of parameters is that the amount of arithmetic performed by mathematicians is extremely small. Many mathematicians dislike all numerical computation and are not particularly expert at it. The territory of arithmetic ends where the two ideas of "variables" and of "algebraic form" commence their sway.

CHAPTER VI

GENERALIZATIONS OF NUMBER

ONE great peculiarity of mathematics is the set of allied ideas which have been invented in connection with the integral numbers from which we started. These ideas may be called extensions or generalizations of number. In the first place there is the idea of fractions. The earliest treatise on arithmetic which we possess was written by an Egyptian priest, named Ahmes, between 1700 B.C. and 1100 B.C., and it is probably a copy of a much older work. It deals largely with the properties of fractions. It appears, therefore, that this concept was developed very early in the history of mathematics. Indeed the subject is a very obvious one. To divide a field into three equal parts, and to take two of the parts, must be a type of operation which had often occurred. Accordingly, we need not be surprised that the men of remote civilizations were familiar with the idea of two-thirds, and with allied notions. Thus as the first generalization of number we place the concept of fractions. The Greeks thought of this subject rather in the form of ratio, so that a Greek would naturally say that a line of two feet in length bears to a line of three feet in length the ratio of 2 to 3. Under the influence of our algebraic notation we would more often say that one line was two-thirds of the other in length, and would think of two-thirds as a numerical multiplier.

In connection with the theory of ratio, or fractions, the

Greeks made a great discovery, which has been the occasion of a large amount of philosophical as well as mathematical thought. They found out the existence of "incommensurable" ratios. They proved, in fact, during the course of their geometrical investigations that, starting with a line of any length, other lines must exist whose lengths do not bear to the original length the ratio of any pair of integers—or, in other words, that lengths exist which are not any exact fraction of the original length.

For example, the diagonal of a square cannot be expressed as any fraction of the side of the same square; in our modern notation the length of the diagonal is $\sqrt{2}$ times the length of the side. But there is no fraction which exactly represents $\sqrt{2}$. We can approximate to $\sqrt{2}$ as closely as we like, but we never exactly reach its value. For example, $\frac{49}{25}$ is just less than 2, and $\frac{9}{4}$ is greater than 2, so that $\sqrt{2}$ lies between $\frac{7}{5}$ and $\frac{3}{2}$. But the best systematic way of approximating to $\sqrt{2}$ in obtaining a series of decimal fractions, each bigger than the last, is by the ordinary method of extracting the square root; thus the series is 1, $\frac{14}{10}$, $\frac{141}{100}$, $\frac{1414}{1000}$, and so on.

Ratios of this sort are called by the Greeks incommensurable. They have excited from the time of the Greeks onwards a great deal of philosophic discussion, and the difficulties connected with them have only recently been cleared up.

We will put the incommensurable ratios with the fractions, and consider the whole set of integral numbers, fractional numbers, and incommensurable numbers as forming one class of numbers which we will call "real numbers." We always think of the real numbers as arranged in order of magnitude, starting from zero and going upwards, and becoming indefinitely larger and larger as we proceed. The real numbers are conveniently represented by points on a line. Let OX be any line bounded at O and stretching away indefi-

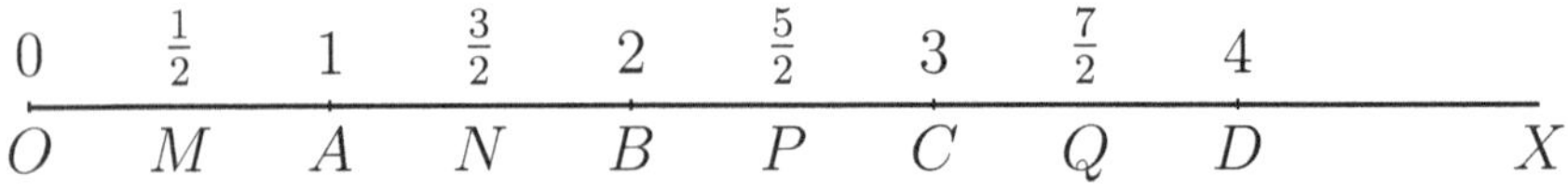

nitely in the direction OX. Take any convenient point, A, on it, so that OA represents the unit length; and divide off lengths AB, BC, CD, and so on, each equal to OA. Then the point O represents the number 0, A the number 1, B the number 2, and so on. In fact the number represented by any point is the measure of its distance from O, in terms of the unit length OA. The points between O and A represent the proper fractions and the incommensurable numbers less than 1; the middle point of OA represents $\frac{1}{2}$, that of AB represents $\frac{3}{2}$, that of BC represents $\frac{5}{2}$, and so on. In this way every point on OX represents some one real number, and every real number is represented by some one point on OX.

The series (or row) of points along OX, starting from O and moving regularly in the direction from O to X, represents the real numbers as arranged in an ascending order of size, starting from zero and continually increasing as we go on.

All this seems simple enough, but even at this stage there are some interesting ideas to be got at by dwelling on these obvious facts. Consider the series of points which represent the integral numbers only, namely, the points, O, A, B, C, D, etc. Here there is a first point O, a definite next point, A, and each point, such as A or B, has one definite immediate predecessor and one definite immediate successor, with the exception of O, which has no predecessor; also the series goes on indefinitely without end. This sort of order is called the type of order of the integers; its essence is the possession of next-door neighbours on either side with the exception of No. 1 in the row. Again consider the integers and fractions together, omitting the points which correspond to the incommensurable ratios. The sort of serial order which we now obtain is quite different. There is a first term O; but no term has any immediate predecessor or immediate successor. This is easily seen to be the case, for between any two fractions we can always find another fraction intermediate in value. One very simple way of doing this is to add the fractions together and to halve the result. For example, between $\frac{2}{3}$ and $\frac{3}{4}$, the fraction $\frac{1}{2}(\frac{2}{3}+\frac{3}{4})$, that is $\frac{17}{24}$, lies; and between $\frac{2}{3}$ and $\frac{17}{24}$ the fraction $\frac{1}{2}(\frac{2}{3}+\frac{17}{24})$, that is $\frac{33}{48}$, lies; and so on indefinitely. Because of this property the series is said to be "compact." There is no end point to the series, which increases indefinitely without limit as we go along the line OX. It would seem at first sight as though the type of series got in this way from the fractions, always including the integers, would be the same as that got from all the real numbers, integers, fractions, and incommensurables taken together, that is, from all the points on the line OX. All that we have hitherto said about the series of fractions applies equally well to the series of all real numbers. But there are important differences which we now proceed to develop. The absence of the incommensurables from the series of fractions leaves an absence of endpoints to certain classes. Thus, consider the incommensurable $\sqrt{2}$. In the series of real numbers this stands between all the numbers whose squares are less than 2, and all the numbers whose squares are greater than 2. But keeping to the series of fractions alone and not thinking of the incommensurables, so that we cannot bring in $\sqrt{2}$, there is no fraction which has the property of dividing off the series into two parts in this way, *i.e.* so that all the members on one side have their squares less than 2, and on the other side greater than 2. Hence in the series of fractions there is a quasi-gap where $\sqrt{2}$ ought to come. This presence of quasi-gaps in the series of fractions may seem a small matter; but any mathematician, who happens to read this, knows that the possible absence of limits or maxima to a class of numbers, which yet does not spread over the whole series of numbers, is no small evil. It is to avoid this difficulty that recourse is had to the incommensurables, so as to obtain a complete series with no gaps.

There is another even more fundamental difference between the two series. We can rearrange the fractions in a series like that of the integers, that is, with a first term, and such that each term has an immediate successor and (except the first term) an immediate predecessor. We can show how this can be done. Let every term in the series of fractions and integers be written in the fractional form by writing $\frac{1}{1}$ for 1, $\frac{2}{1}$ for 2, and so on for all the integers, excluding 0. Also for the moment we will reckon fractions which are equal in value but not reduced to their lowest terms as distinct; so that, for example, until further notice $\frac{2}{3}, \frac{4}{6}, \frac{6}{9}, \frac{8}{12}$, etc., are all reckoned as distinct. Now group the fractions into classes by adding together the numerator and denominator of each term. For the sake of brevity call this sum of the numerator and denominator of a fraction its index. Thus 7 is the index of $\frac{4}{3}$, and also of $\frac{3}{4}$, and of $\frac{2}{5}$. Let the fractions in each class be all fractions which have some specified index, which may therefore also be called the class index. Now arrange these classes in the order of magnitude of their indices. The first class has the index 2, and its only member is $\frac{1}{1}$; the second class has the index 3, and its members are $\frac{1}{2}$ and $\frac{2}{1}$; the third class has the index 4, and its members are $\frac{1}{3}, \frac{2}{2}, \frac{3}{1}$; the fourth class has the index 5, and its members are $\frac{1}{4}, \frac{2}{3}, \frac{3}{2}, \frac{4}{1}$; and so on. It is easy to see that the number of members (still including fractions not in their lowest terms) belonging to any class is one less than its index. Also the members of any one class can be arranged in order by taking the first member to be the fraction with numerator 1, the second member to have the numerator 2, and so on, up to $(n-1)$ where n is the index. Thus for the class of index n, the members appear in the order

$$\frac{1}{n-1}, \quad \frac{2}{n-2}, \quad \frac{3}{n-3}, \ \ldots, \quad \frac{n-1}{1}.$$

The members of the first four classes have in fact been mentioned in this order. Thus the whole set of fractions have now been arranged in an order like that of the integers. It runs thus

$$\frac{1}{1}, \frac{1}{2}, \frac{2}{1}, \frac{1}{3}, \left[\frac{2}{2}\right], \frac{3}{1}, \frac{1}{4}, \frac{2}{3}, \frac{3}{2}, \frac{4}{1}, \ldots,$$
$$\frac{n-1}{1}, \frac{1}{n-1}, \frac{2}{n-2}, \frac{3}{n-3}, \ldots, \frac{n-1}{1}, \frac{1}{n},$$

and so on.

Now we can get rid of all repetitions of fractions of the same value by simply striking them out whenever they appear after their first occurrence. In the few initial terms written down above, $\frac{2}{2}$ which is enclosed above in square brackets is the only fraction not in its lowest terms. It has occurred before as $\frac{1}{1}$. Thus this must be struck out. But the series is still left with the same properties, namely, (a) there is a first term, (b) each term has next-door neighbours, (c) the series goes on without end.

It can be proved that it is not possible to arrange the whole series of real numbers in this way. This curious fact was discovered by Georg Cantor, a German mathematician still living; it is of the utmost importance in the philosophy of mathematical ideas. We are here in fact touching on the fringe of the great problems of the meaning of continuity and of infinity.

Another extension of number comes from the introduction of the idea of what has been variously named an operation or a step, names which are respectively appropriate from slightly different points of view. We will start with a particular case. Consider the statement $2 + 3 = 5$. We add 3 to 2 and obtain 5. Think of the operation of adding 3: let this be denoted by $+3$. Again $4 - 3 = 1$. Think of the operation of subtracting 3: let this be denoted by -3. Thus instead of considering the real numbers in themselves, we consider the *operations* of adding or subtracting them: instead of $\sqrt{2}$, we consider $+\sqrt{2}$ and $-\sqrt{2}$, namely the operations of adding $\sqrt{2}$ and of subtracting $\sqrt{2}$. Then we can add these operations, of course in a different sense of addition to that in which we add numbers. The sum of two operations is the single operation which has the same effect as the two operations applied successively. In what order are the two operations to be applied? The answer is that it is indifferent, since for example

$$2 + 3 + 1 = 2 + 1 + 3;$$

so that the addition of the steps $+3$ and $+1$ is commutative.

Mathematicians have a habit, which is puzzling to those engaged in tracing out meanings, but is very convenient in practice, of using the same symbol in different though allied senses. The one essential requisite for a symbol in their eyes is that, whatever its possible varieties of meaning, the formal laws for its use shall always be the same. In accordance with this habit the addition of operations is denoted by + as well as the addition of numbers. Accordingly we can write

$$(+3) + (+1) = +4;$$

where the middle + on the left-hand side denotes the addition of the operations +3 and +1. But, furthermore, we need not be so very pedantic in our symbolism, except in the rare instances when we are directly tracing meanings; thus we always drop the first + of a line and the brackets, and never write two + signs running. So the above equation becomes

$$3 + 1 = 4,$$

which we interpret as simple numerical addition, or as the more elaborate addition of operations which is fully expressed in the previous way of writing the equation, or lastly as expressing the result of applying the operation +1 to the number 3 and obtaining the number 4. Any interpretation which is possible is always correct. But the only interpretation which is always possible, under certain conditions, is that of operations. The other interpretations often give nonsensical results.

This leads us at once to a question, which must have been rising insistently in the reader's mind: What is the use of all this elaboration? At this point our friend, the practical man, will surely step in and insist on sweeping away all these silly cobwebs of the brain. The answer is that what the mathematician is seeking is Generality. This is an idea worthy to be placed beside the notions of the Variable and of Form so far as concerns its importance in governing mathematical procedure. Any limitation whatsoever upon the generality of theorems, or of proofs, or of interpretation is abhorrent to the mathematical instinct. These three notions, of the variable, of form, and of generality, compose a sort of mathematical trinity which preside over the whole subject. They all really spring from the same root, namely from the abstract nature of the science.

Let us see how generality is gained by the introduction of this idea of operations. Take the equation $x + 1 = 3$; the solution is $x = 2$. Here we can interpret our symbols as mere numbers, and the recourse to "operations" is entirely unnecessary. But, if x is a mere number, the equation $x + 3 = 1$ is nonsense. For x should be the number of things which remain when you have taken 3 things away from 1 thing; and no such procedure is possible. At this point our idea of algebraic form steps in, itself only generalization under another aspect. We consider, therefore, the general equation of the same form as $x + 1 = 3$. This equation is $x + a = b$, and its solution is $x = b - a$. Here our difficulties become acute; for this form can only be used for the numerical interpretation so long as b is greater than a, and we cannot say without qualification that a and b may be any constants. In other words we have introduced a limitation on the variability of the "constants" a and b, which we must drag like a chain throughout all our reasoning. Really prolonged mathematical investigations would be impossible under such conditions. Every equation would at last be buried under a pile of limitations. But if we now interpret our symbols as "operations," all limitation vanishes like magic. The equation $x + 1 = 3$ gives $x = +2$, the equation $x + 3 = 1$ gives $x = -2$, the equation $x + a = b$ gives $x = b - a$ which is an operation of addition or subtraction as the case may be. We need never decide whether $b - a$ represents the operation of addition or of subtraction, for the rules of procedure with the symbols are the same in either case.

It does not fall within the plan of this work to write a detailed chapter of elementary algebra. Our object is merely to make plain the fundamental ideas which guide the formation of the science. Accordingly we do not further explain the detailed rules by which the "positive and negative numbers" are multiplied and otherwise combined. We have explained above that positive and negative numbers are operations. They have also been called "steps." Thus $+3$ is the step by which we go from 2 to 5, and -3 is the step backwards by which we go from 5 to 2. Consider the line OX divided in the way explained in the earlier part of the chapter, so that its points represent numbers. Then $+2$ is the step from O

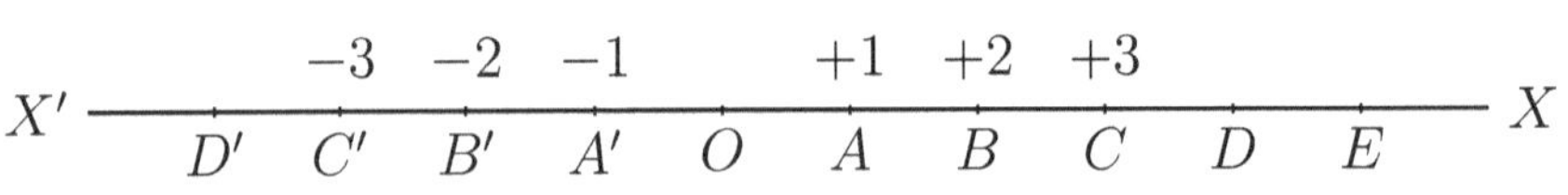

to B, or from A to C, or (if the divisions are taken backwards along OX') from C' to A', or from D' to B', and so on. Similarly -2 is the step from O to B', or from B' to D', or from B to O, or from C to A.

We may consider the point which is reached by a step from O, as representative of that step. Thus A represents $+1$, B represents $+2$, A' represents -1, B' represents -2, and so on. It will be noted that, whereas previously with the mere "unsigned" real numbers the points on one side of O only, namely along OX, were representative of numbers, now with steps every point on the whole line stretching on both sides of O is representative of a step. This is a pictorial representation of the superior generality introduced by the positive and negative numbers, namely the operations or steps. These "signed" numbers are also particular cases of what have been called vectors (from the Latin *veho*, I draw or carry). For we may think of a particle as carried from O to A, or from A to B.

In suggesting a few pages ago that the practical man would object to the subtlety involved by the introduction of the positive and negative numbers, we were libelling that excellent individual. For in truth we are on the scene of one of his greatest triumphs. If the truth must be confessed, it was the practical man himself who first employed the actual symbols $+$ and $-$. Their origin is not very certain, but it seems most probable that they arose from the marks chalked on chests of goods in German warehouses, to denote excess or defect from some standard weight. The earliest notice of them occurs in a book published at Leipzig, in A.D. 1489. They seem first to have been employed in mathematics by a German mathematician, Stifel, in a book published at Nuremburg in 1544 A.D. But then it is only recently that the Germans have come to be looked on as emphatically a practical nation. There is an old epigram which assigns the empire of the sea to the English, of the land to the French, and of the clouds to the Germans. Surely it was from the clouds that the Germans fetched + and −; the ideas which these symbols have generated are much too important for the welfare of humanity to have come from the sea or from the land.

The possibilities of application of the positive and negative numbers are very obvious. If lengths in one direction are represented by positive numbers, those in the opposite direction are represented by negative numbers. If a velocity in one direction is positive, that in the opposite direction is negative. If a rotation round a dial in the opposite direction to the hands of a clock (anti-clockwise) is positive, that in the clockwise direction is negative. If a balance at the bank is positive, an overdraft is negative. If vitreous electrification is positive, resinous electrification is negative. Indeed, in this latter case, the terms positive electrification and negative electrification, considered as mere names, have practically driven out the other terms. An endless series of examples could be given. The idea of positive and negative numbers has been practically the most successful of mathematical subtleties.

CHAPTER VII

IMAGINARY NUMBERS

If the mathematical ideas dealt with in the last chapter have been a popular success, those of the present chapter have excited almost as much general attention. But their success has been of a different character, it has been what the French term a *succès de scandale.* Not only the practical man, but also men of letters and philosophers have expressed their bewilderment at the devotion of mathematicians to mysterious entities which by their very name are confessed to be imaginary. At this point it may be useful to observe that a certain type of intellect is always worrying itself and others by discussion as to the applicability of technical terms. Are the incommensurable numbers properly called numbers? Are the positive and negative numbers really numbers? Are the imaginary numbers imaginary, and are they numbers?—are types of such futile questions. Now, it cannot be too clearly understood that, in science, technical terms are names arbitrarily assigned, like Christian names to children. There can be no question of the names being right or wrong. They may be judicious or injudicious; for they can sometimes be so arranged as to be easy to remember, or so as to suggest relevant and important ideas. But the essential principle involved was quite clearly enunciated in Wonderland to Alice by Humpty Dumpty, when he told her, à propos of his use of words, "I pay them extra and make them mean what I like." So we will not bother as to whether imaginary numbers are imaginary, or as to whether they are numbers, but will take the phrase as the arbitrary name of a certain mathematical idea, which we will now endeavour to make plain.

The origin of the conception is in every way similar to that of the positive and negative numbers. In exactly the same way it is due to the three great mathematical ideas of the variable, of algebraic form, and of generalization. The positive and negative numbers arose from the consideration of equations like $x + 1 = 3$, $x + 3 = 1$, and the general form $x + a = b$. Similarly the origin of imaginary numbers is due to equations like $x^2 + 1 = 3$, $x^2 + 3 = 1$, and $x^2 + a = b$. Exactly the same process is gone through. The equation $x^2 + 1 = 3$ becomes $x^2 = 2$, and this has two solutions, either $x = +\sqrt{2}$, or $x = -\sqrt{2}$. The statement that there are these alternative solutions is usually written $x = \pm\sqrt{2}$. So far all is plain sailing, as it was in the previous case. But now an analogous difficulty arises. For the equation $x^2 + 3 = 1$ gives $x^2 = -2$ and there is no positive or negative number which, when multiplied by itself, will give a negative square. Hence, if our symbols are to mean the ordinary positive or negative numbers, there is no solution to $x^2 = -2$, and the equation is in fact nonsense. Thus, finally taking the general form $x^2+a = b$, we find the pair of solutions $x = \pm\sqrt{(b-a)}$, when, and only when, b is not less than a. Accordingly we cannot say unrestrictedly that the "constants" a and b may be any numbers, that is, the "constants" a and b are not, as they ought to be, independent unrestricted "variables"; and so again a host of limitations and restrictions will accumulate round our work as we proceed.

The same task as before therefore awaits us: we must give a new interpretation to our symbols, so that the solutions $\pm\sqrt{(b-a)}$ for the equation $x^2 + a = b$ always have meaning. In other words, we require an interpretation of the symbols so that $\sqrt{a}$ always has meaning whether a be positive or negative. Of course, the interpretation must be such that all the ordinary formal laws for addition, subtraction, multiplication, and division hold good; and also it must not interfere with the generality which we have attained by the use of the positive and negative numbers. In fact, it must in a sense include them as special cases. When a is negative we may write $-c^2$ for it, so that c^2 is positive. Then

$$\begin{aligned}\sqrt{a} &= \sqrt{(-c^2)} = \sqrt{\{(-1) \times c^2\}} \\ &= \sqrt{(-1)}\sqrt{c^2} = c\sqrt{(-1)}.\end{aligned}$$

Hence, if we can so interpret our symbols that $\sqrt{(-1)}$ has a meaning, we have attained our object. Thus $\sqrt{(-1)}$ has come to be looked on as the head and forefront of all the imaginary quantities.

This business of finding an interpretation for $\sqrt{(-1)}$ is a much tougher job than the analogous one of interpreting -1. In fact, while the easier problem was solved almost instinctively as soon as it arose, it at first hardly occurred, even to the greatest mathematicians, that here a problem existed which was perhaps capable of solution. Equations like $x^2 = -3$, when they arose, were simply ruled aside as nonsense.

However, it came to be gradually perceived during the eighteenth century, and even earlier, how very convenient it would be if an interpretation could be assigned to these nonsensical symbols. Formal reasoning with these symbols was gone through, merely assuming that they obeyed the ordinary algebraic laws of transformation; and it was seen that a whole world of interesting results could be attained, if only these symbols might legitimately be used. Many mathematicians were not then very clear as to the logic of their procedure, and an idea gained ground that, in some mysterious way, symbols which mean nothing can by appropriate manipulation yield valid proofs of propositions. Nothing can be more mistaken. A symbol which has not been properly defined is not a symbol at all. It is merely a blot of ink on paper which has an easily recognized shape. Nothing can be proved by a succession of blots, except the existence of a bad pen or a careless writer. It was during this epoch that the epithet "imaginary" came to be applied to $\sqrt{(-1)}$. What these mathematicians had really succeeded in proving were a series of hypothetical propositions, of which this is the blank form: If interpretations exist for $\sqrt{(-1)}$ and for the addition, subtraction, multiplication, and division of $\sqrt{(-1)}$ which make the ordinary algebraic rules (*e.g.* $x + y = y + x$, etc.) to be satisfied, then such and such results follows. It was natural that the mathematicians should not always appreciate the big "If," which ought to have preceded the statements of their results.

As may be expected the interpretation, when found, was a much more elaborate affair than that of the negative numbers and the reader's attention must be asked for some careful preliminary explanation. We have already come across the representation of a point by two numbers. By the aid of the positive and negative numbers we can now represent the

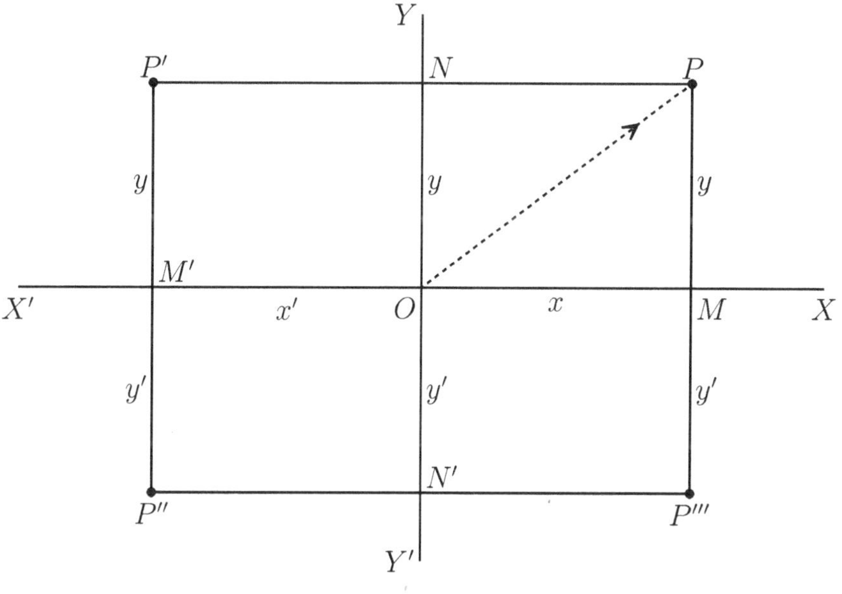

Fig. 8.

position of any point in a plane by a pair of such numbers. Thus we take the pair of straight lines XOX' and YOY', at right angles, as the "axes" from which we start all our measurements. Lengths measured along OX and OY are positive, and measured backwards along OX' and OY' are negative. Suppose that a pair of numbers, written in order, *e.g.* $(+3, +1)$, so that there is a first number ($+3$ in the above example), and a second number ($+1$ in the above example), represents measurements from O along XOX' for the first number, and along YOY' for the second number. Thus (*cf.* fig. 9) in $(+3, +1)$ a length of 3 units is to be measured along XOX' in the positive direction, that is from O towards X, and a length $+1$ measured along YOY' in the positive direction, that is from O towards Y. Similarly in $(-3, +1)$ the length of 3 units is to be measured from O towards X', and of 1 unit from towards Y. Also in $(-3, -1)$ the two lengths are to be measured along OX' and OY' respectively, and in $(+3, -1)$ along OX and OY' respectively. Let us for the moment call such a pair of numbers an "ordered couple." Then, from the two numbers 1 and 3, eight ordered couples can be generated, namely

$$(+1, +3),\ (-1, +3),\ (-1, -3),\ (+1, -3),$$
$$(+3, +1),\ (-3, +1),\ (-3, -1),\ (+3, -1).$$

Each of these eight "ordered couples" directs a process of measurement along XOX' and YOY' which is different from that directed by any of the others.

The processes of measurement represented by the last four ordered couples, mentioned above, are given pictorially in the figure. The lengths OM and ON together correspond to $(+3, +1)$, the lengths OM' and ON together correspond to $(-3, +1)$, OM' and ON' together to $(-3, -1)$, and OM and ON' together to $(+3, -1)$. But by completing the various rectangles, it is easy to see that the point P completely determines and is determined by the ordered couple $(+3, +1)$, the point P' by $(-3, +1)$, the point P'' by

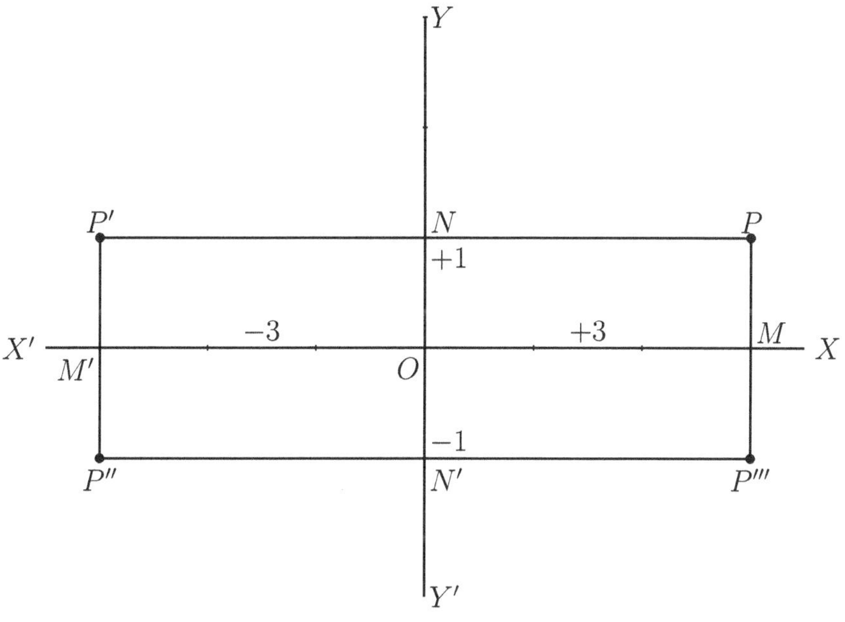

Fig. 9.

$(-3, -1)$, and the point P''' by $(+3, -1)$. More generally in the previous figure (8), the point P corresponds to the ordered couple (x, y), where x and y in the figure are both assumed to be positive, the point P' corresponds to (x', y), where x' in the figure is assumed to be negative, P'' to $(x'y')$, and P''' to (x, y'). Thus an ordered couple (x, y), where x and y are any positive or negative numbers, and the corresponding point reciprocally determine each other. It is convenient to introduce some names at this juncture. In the ordered couple (x, y) the first number x is called the "abscissa" of the corresponding point, and the second number y is called the "ordinate" of the point, and the two numbers together are called the "coordinates" of the point. The idea of determining the position of a point by its "coordinates" was by no means new when the theory of "imaginaries" was being formed. It was due to Descartes, the great French mathematician and philosopher, and appears in his *Discours* published at Leyden in 1637 A.D. The idea of the ordered couple as a thing on its own account is of later growth and is the outcome of the efforts to interpret imaginaries in the most abstract way possible.

It may be noticed as a further illustration of this idea of the ordered couple, that the point M in fig. 9 is the couple $(+3, 0)$, the point N is the couple $(0, +1)$, the point M' the couple $(-3, 0)$, the point N' the couple $(0, -1)$, the point O the couple $(0, 0)$.

Another way of representing the ordered couple (x, y) is to think of it as representing the dotted line OP (*cf.* fig. 8), rather than the point P. Thus the ordered couple represents a line drawn from an "origin," O, of a certain length and in a certain direction. The line OP may be called the vector line from O to P, or the step from O to P. We see, therefore, that we have in this chapter only extended the interpretation which we gave formerly of the positive and negative numbers. This method of representation by vectors is very useful when we consider the meaning to be assigned to the operations of the addition and multiplication of ordered couples.

We will now go on to this question, and ask what meaning we shall find it convenient to assign to the addition of the two ordered couples (x, y) and (x', y'). The interpretation must, (a) make the result of addition to be another ordered couple, (b) make the operation commutative so that $(x, y) + (x', y') = (x', y') + (x, y)$, (c) make the operation associative so that

$$\{(x, y) + (x', y')\} + (u, v) = (x, y) + \{(x', y') + (u, v)\},$$

(d) make the result of subtraction unique, so that when we seek to determine the unknown ordered couple (x, y) so as to satisfy the equation

$$(x, y) + (a, b) = (c, d),$$

there is one and only one answer which we can represent by

$$(x, y) = (c, d) - (a, b).$$

All these requisites are satisfied by taking $(x, y) + (x', y')$ to mean the ordered couple $(x + x', y + y')$. Accordingly by definition we put

$$(x, y) + (x', y') = (x + x', y + y').$$

Notice that here we have adopted the mathematical habit of using the same symbol + in different senses. The + on the left-hand side of the equation has the new meaning of + which we are just defining; while the two +'s on the right-hand side have the meaning of the addition of positive and negative numbers (operations) which was defined in the last chapter. No practical confusion arises from this double use.

As examples of addition we have

$$(+3, +1) + (+2, +6) = (+5, +7),$$
$$(+3, -1) + (-2, -6) = (+1, -7),$$
$$(+3, +1) + (-3, -1) = (0, 0).$$

The meaning of subtraction is now settled for us. We find that

$$(x, y) - (u, v) = (x - u, y - v).$$

Thus

$$(+3, +2) - (+1, +1) = (+2, +1),$$

and

$$(+1, -2) - (+2, -4) = (-1, +2),$$

and

$$(-1, -2) - (+2, +3) = (-3, -5).$$

It is easy to see that

$$(x, y) - (u, v) = (x, y) + (-u, -v).$$

Also

$$(x, y) - (x, y) = (0, 0).$$

Hence $(0, 0)$ is to be looked on as the zero ordered couple. For example

$$(x, y) + (0, 0) = (x, y).$$

The pictorial representation of the addition of ordered couples is surprisingly easy.

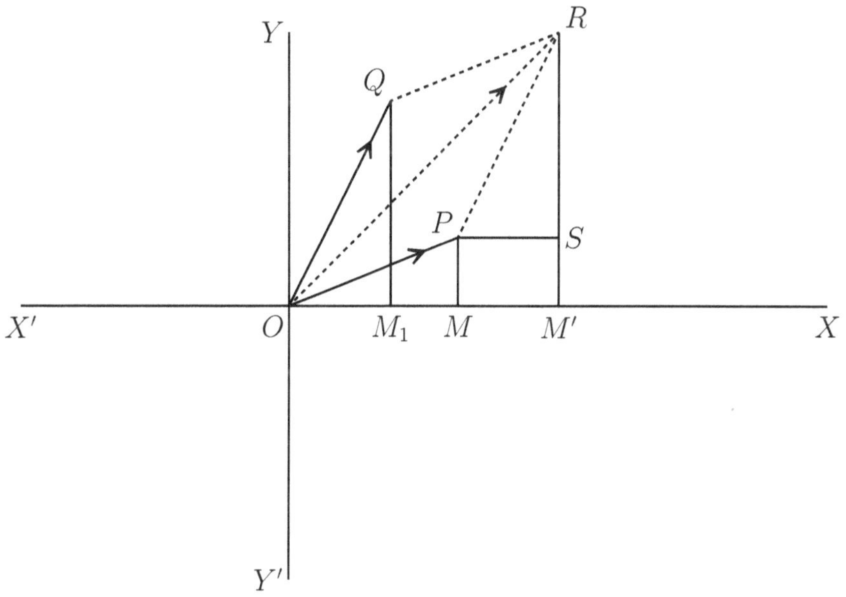

Fig. 10.

Let OP represent (x, y) so that $OM = x$ and $PM = y$; let OQ represent (x_1, y_1) so that $OM_1 = x_1$ and $QM_1 = y_1$. Complete the parallelogram $OPRQ$ by the dotted lines PR and QR, then the diagonal OR is the ordered couple $(x + x_1, y + y_1)$. For draw PS parallel to OX; then evidently the triangles OQM_1 and PRS are in all respects equal. Hence $MM' = PS = x_1$, and $RS = QM_1$ and therefore

$$OM' = OM + MM' = x + x_1,$$
$$RM' = SM' + RS = y + y_1.$$

Thus OR represents the ordered couple as required. This figure can also be drawn with OP and OQ in other quadrants.

It is at once obvious that we have here come back to the parallelogram law, which was mentioned in Chapter VI., on the laws of motion, as applying to velocities and forces. It will be remembered that, if OP and OQ represent two velocities, a particle is said to be moving with a velocity equal to the two velocities added together if it be moving with the velocity OR. In other words OR is said to be the resultant of the two velocities OP and OQ. Again forces acting at a point of a body can be represented by lines just as velocities can be; and the same parallelogram law holds, namely, that the resultant of the two forces OP and OQ is the force represented by the diagonal OR. It follows that we can look on an ordered couple as representing a velocity or a force, and the rule which we have just given for the addition of ordered couples then represents the fundamental laws of mechanics for the addition of forces and velocities. One of the most fascinating characteristics of mathematics is the surprising way in which the ideas and results of different parts of the subject dovetail into each other. During the discussions of this and the previous chapter we have been guided merely by the most abstract of pure mathematical considerations; and yet at the end of them we have been led back to the most fundamental of all the laws of nature, laws which have to be in the mind of every engineer as he designs an engine, and of every naval architect as he calculates the stability of a ship. It is no paradox to say that in our most theoretical moods we may be nearest to our most practical applications.

CHAPTER VIII

IMAGINARY NUMBERS (*Continued*)

THE definition of the multiplication of ordered couples is guided by exactly the same considerations as is that of their addition. The interpretation of multiplication must be such that

(α) the result is another ordered couple,

(β) the operation is commutative, so that

$$(x, y) \times (x', y') = (x', y') \times (x, y),$$

(γ) the operation is associative, so that

$$\{(x, y) \times (x', y')\} \times (u, v) = (x, y) \times \{(x', y') \times (u, v)\},$$

(δ) must make the result of division unique [with an exception for the case of the zero couple $(0, 0)$], so that when we seek to determine the unknown couple (x, y) so as to satisfy the equation

$$(x, y) \times (a, b) = (c, d),$$

there is one and only one answer, which we can represent by

$$(x, y) = (c, d) \div (a, b), \quad \text{or by} \quad (x, y) = \frac{(c, d)}{(a, b)}.$$

(ϵ) Furthermore the law involving both addition and multiplication, called the distributive law, must be satisfied, namely

$$(x, y) \times \{(a, b) + (c, d)\}$$
$$= \{(x, y) \times (a, b)\} + \{(x, y) \times (c, d)\}.$$

All these conditions (α), (β), (γ), (δ), (ϵ) can be satisfied by an interpretation which, though it looks complicated at first, is capable of a simple geometrical interpretation.

By definition we put

$$(A) \qquad (x, y) \times (x', y') = \{(xx' - yy'), (xy' + x'y)\}.$$

This is the definition of the meaning of the symbol $\times$ when it is written between two ordered couples. It follows evidently from this definition that the result of multiplication is another ordered couple, and that the value of the right-hand side of equation (A) is not altered by simultaneously interchanging x with x', and y with y'. Hence conditions (α) and (β) are evidently satisfied. The proof of the satisfaction of (γ), (δ), (ϵ) is equally easy when we have given the geometrical interpretation, which we will proceed to do in a moment. But before doing this it will be interesting to pause and see whether we have attained the object for which all this elaboration was initiated.

We came across equations of the form $x^2 = -3$, to which no solutions could be assigned in terms of positive and negative real numbers. We then found that all our difficulties would vanish if we could interpret the equation $x^2 = -1$, *i.e.*, if we could so define $\sqrt{(-1)}$ that $\sqrt{(-1)} \times \sqrt{(-1)} = -1$.

Now let us consider the three special ordered couples* $(0, 0)$, $(1, 0)$, and $(0, 1)$.

We have already proved that

$$(x, y) + (0, 0) = (x, y).$$

Furthermore we now have

$$(x, y) \times (0, 0) = (0, 0).$$

Hence both for addition and for multiplication the couple $(0, 0)$ plays the part of zero in elementary arithmetic and algebra; compare the above equations with $x + 0 = x$, and $x \times 0 = 0$.

Again consider $(1, 0)$: this plays the part of 1 in elementary arithmetic and algebra. In these elementary sciences the special characteristic of 1 is that $x \times 1 = x$, for all values of x. Now by our law of multiplication

$$(x, y) \times (1, 0) = \{(x - 0), (y + 0)\} = (x, y).$$

Thus $(1, 0)$ is the unit couple.

Finally consider $(0, 1)$: this will interpret for us the symbol $\sqrt{(-1)}$. The symbol must therefore possess the characteristic property that $\sqrt{(-1)} \times \sqrt{(-1)} = -1$. Now by the law of multiplication for ordered couples

$$(0, 1) \times (0, 1) = \{(0 - 1), (0 + 0)\} = (-1, 0).$$

---

*For the future we follow the custom of omitting the + sign wherever possible, thus $(1, 0)$ stands for $(+1, 0)$ and $(0, 1)$ for $(0, +1)$.

But $(1, 0)$ is the unit couple, and $(-1, 0)$ is the negative unit couple; so that $(0, 1)$ has the desired property. There are, however, two roots of -1 to be provided for, namely $\pm\sqrt{(-1)}$. Consider $(0, -1)$; here again remembering that $(-1)^2 = 1$, we find, $(0, -1) \times (0, -1) = (-1, 0)$.

Thus $(0, -1)$ is the other square root of -1. Accordingly the ordered couples $(0, 1)$ and $(0, -1)$ are the interpretations of $\pm\sqrt{(-1)}$ in terms of ordered couples. But which corresponds to which? Does $(0, 1)$ correspond to $+\sqrt{(-1)}$ and $(0, -1)$ to $-\sqrt{(-1)}$, or $(0, 1)$ to $-\sqrt{(-1)}$, and $(0, -1)$ to $+\sqrt{(-1)}$? The answer is that it is perfectly indifferent which symbolism we adopt.

The ordered couples can be divided into three types, (i) the "complex imaginary" type (x, y), in which neither x nor y is zero; (ii) the "real" type $(x, 0)$; (iii) the "pure imaginary" type $(0, y)$. Let us consider the relations of these types to each other. First multiply together the "complex imaginary" couple (x, y) and the "real" couple $(a, 0)$, we find

$$(a, 0) \times (x, y) = (ax, ay).$$

Thus the effect is merely to multiply each term of the couple (x, y) by the positive or negative real number a.

Secondly, multiply together the "complex imaginary" couple (x, y) and the "pure imaginary" couple $(0, b)$, we find

$$(0, b) \times (x, y) = (-by, bx).$$

Here the effect is more complicated, and is best comprehended in the geometrical interpretation to which we proceed after noting three yet more special cases.

Thirdly, we multiply the "real" couple $(a, 0)$ by the imaginary $(0, b)$ and obtain

$$(a, 0) \times (0, b) = (0, ab).$$

Fourthly, we multiply the two "real" couples $(a, 0)$ and $(a', 0)$ and obtain

$$(a, 0) \times (a', 0) = (aa', 0).$$

Fifthly, we multiply the two "imaginary couples" $(0, b)$ and $(0, b')$ and obtain

$$(0, b) \times (0, b') = (-bb', 0).$$

We now turn to the geometrical interpretation, beginning first with some special cases. Take the couples $(1, 3)$ and $(2, 0)$ and consider the equation

$$(2, 0) \times (1, 3) = (2, 6).$$

In the diagram (fig. 11) the vector OP represents $(1, 3)$, and the vector ON represents $(2, 0)$, and the vector OQ represents $(2, 6)$. Thus the product $(2, 0) \times (1, 3)$ is found geometrically by taking the length of the vector OQ to be the product of the lengths of the vectors OP and ON, and (in this case) by producing OP to Q to be of the required length. Again, consider the product $(0, 2) \times (1, 3)$, we have

$$(0, 2) \times (1, 3) = (-6, 2).$$

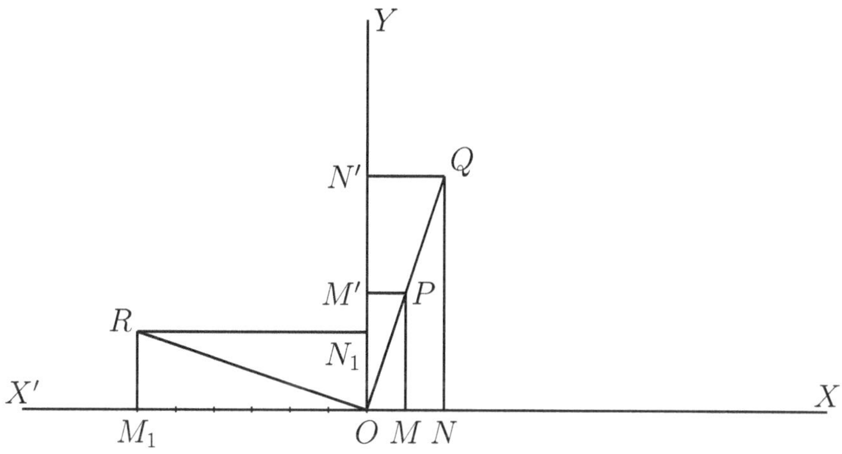

Fig. 11.

The vector ON_1, corresponds to $(0, 2)$ and the vector OR to $(-6, 2)$. Thus OR which represents the new product is at right angles to OQ and of the same length. Notice that we have the same law regulating the length of OQ as in the previous case, namely, that its length is the product of the lengths of the two vectors which are multiplied together; but now that we have ON_1 along the "ordinate" axis OY, instead of ON along the "abscissa" axis OX, the direction of OP has been turned through a right-angle.

Hitherto in these examples of multiplication we have looked on the vector OP as modified by the vectors ON and ON_1. We shall get a clue to the general law for the direction by inverting the way of thought, and by thinking of the vectors ON and ON_1 as modified by the vector OP. The law for the length remains unaffected; the resultant length is the length of the product of the two vectors. The new direction for the enlarged ON (*i.e.* OQ) is found by rotating it in the (anti-clockwise) direction of rotation from OX towards OY through an angle equal to the angle XOP: it is an accident of this particular case that this rotation makes OQ lie along the line OP. Again consider the product of ON_1 and OP; the new direction for the enlarged ON_1 (*i.e.* OR) is found by rotating ON in the anti-clockwise direction of rotation through an angle equal to the angle XOP, namely, the angle N_1OR is equal to the angle XOP.

The general rule for the geometrical representation of multiplication can now be enunciated thus:

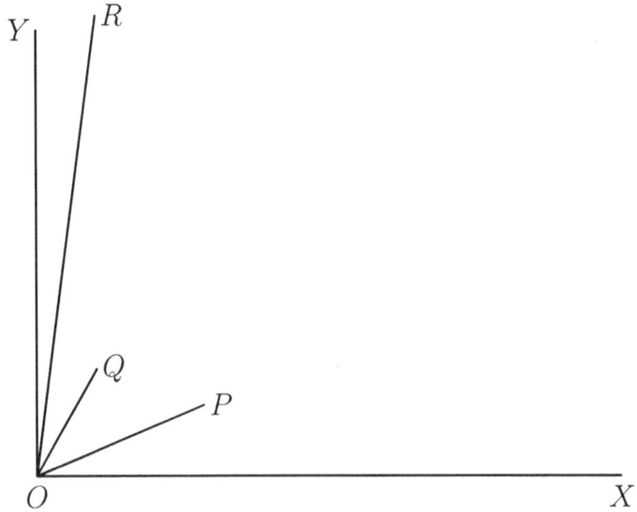

Fig. 12.

The product of the two vectors OP and OQ is a vector OR, whose length is the product of the lengths of OP and OQ and whose direction OR is such that the angle XOR is equal to the sum of the angles XOP and XOQ.

Hence we can conceive the vector OP as making the vector OQ rotate through an angle XOP (*i.e.* the angle $QOR =$ the angle XOP), or the vector OQ as making the vector OP rotate through the angle XOQ (*i.e.*the angle $POR =$ the angle XOQ).

We do not prove this general law, as we should thereby be led into more technical processes of mathematics than falls within the design of this book. But now we can immediately see that the associative law [numbered (γ) above] for multiplication is satisfied. Consider first the length of the resultant vector; this is got by the ordinary process of multiplication for real numbers; and thus the associative law holds for it.

Again, the direction of the resultant vector is got by the mere addition of angles, and the associative law holds for this process also.

So much for multiplication. We have now rapidly indicated, by considering addition and multiplication, how an algebra or "calculus" of vectors in one plane can be constructed, which is such that any two vectors in the plane can be added, or subtracted, and can be multiplied, or divided one by the other.

We have not considered the technical details of all these processes because it would lead us too far into mathematical details; but we have shown the general mode of procedure. When we are interpreting our algebraic symbols in this way, we are said to be employing "imaginary quantities" or "complex quantities." These terms are mere details, and we have far too much to think about to stop to enquire whether they are or are not very happily chosen.

The nett result of our investigations is that any equations like $x+3=2$ or $(x+3)^2=-2$ can now always be interpreted into terms of vectors, and solutions found for them. In seeking for such interpretations it is well to note that 3 becomes $(3,0)$, and -2 becomes $(-2,0)$, and x becomes the "unknown" couple (u,v): so the two equations become respectively $(u,v)+(3,0)=(2,0)$, and $\{(u,v)+(3,0)\}^2=(-2,0)$.

We have now completely solved the initial difficulties which caught our eye as soon as we considered even the elements of algebra. The science as it emerges from the solution is much more complex in ideas than that with which we started. We have, in fact, created a new and entirely different science, which will serve all the purposes for which the old science was invented and many more in addition. But, before we can congratulate ourselves on this result to our labours, we must allay a suspicion which ought by this time to have arisen in the mind of the student. The question which the reader ought to be asking himself is: Where is all this invention of new interpretations going to end? It is true that we have succeeded in interpreting algebra so as always to be able to solve a quadratic equation like $x^2-2x+4=0$; but there are an endless number of other equations, for example, $x^3 - 2x + 4 = 0$, $x^4 + x^3 + 2 = 0$, and so on without limit. Have we got to make a new science whenever a new equation appears?

Now, if this were the case, the whole of our preceding investigations, though to some minds they might be amusing, would in truth be of very trifling importance. But the great fact, which has made modern analysis possible, is that, by the aid of this calculus of vectors, every formula which arises can receive its proper interpretation; and the "unknown" quantity in every equation can be shown to indicate some vector. Thus the science is now complete in itself as far as its fundamental ideas are concerned. It was receiving its final form about the same time as when the steam engine was being perfected, and will remain a great and powerful weapon for the achievement of the victory of thought over things when curious specimens of that machine repose in museums in company with the helmets and breastplates of a slightly earlier epoch.

CHAPTER IX

COORDINATE GEOMETRY

THE methods and ideas of coordinate geometry have already been employed in the previous chapters. It is now time for us to consider them more closely for their own sake; and in doing so we shall strengthen our hold on other ideas to which we have attained. In the present and succeeding chapters we will go back to the idea of the positive and negative real numbers and will ignore the imaginaries which were introduced in the last two chapters.

We have been perpetually using the idea that, by taking two axes, XOX' and YOY', in a plane, any point P in that plane can be determined in position by a pair of positive or negative numbers x and y, where (*cf.* fig. 13) x is the length OM and y is the length PM. This conception, simple as it looks, is the main idea of the great subject of coordinate geometry. Its discovery marks a momentous epoch in the history of mathematical thought. It is due (as has been already said) to the philosopher Descartes, and occurred to him as an important mathematical method one morning as he lay in bed. Philosophers, when they have possessed a thorough knowledge of mathematics, have been among those who have enriched the science with some of its best ideas. On the other hand it must be said that, with hardly an exception, all the remarks on mathematics made by those philosophers who have possessed but a slight or hasty and late-acquired

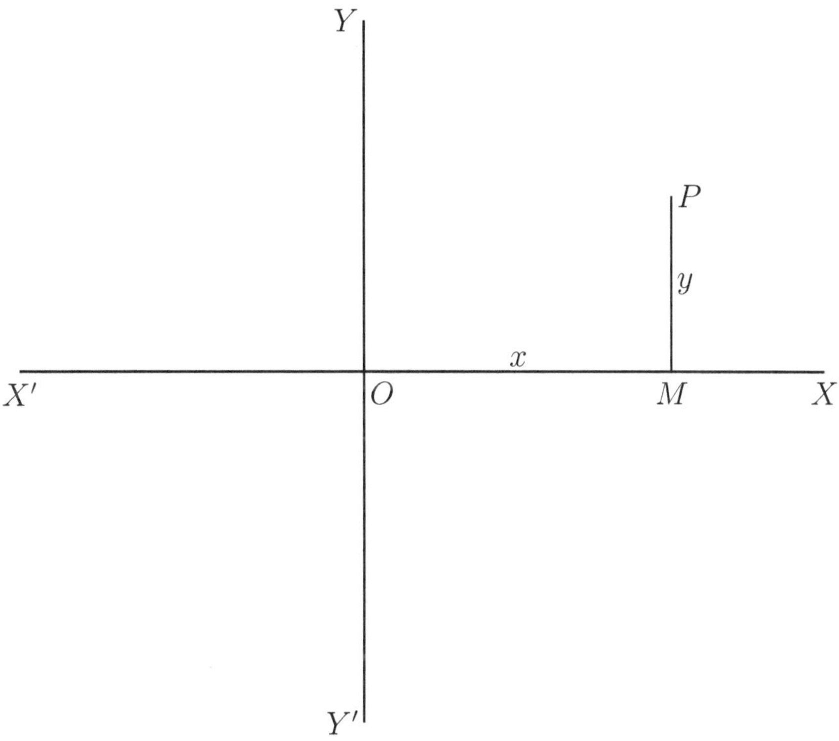

Fig. 13.

knowledge of it are entirely worthless, being either trivial or wrong. The fact is a curious one; since the ultimate ideas of mathematics seem, after all, to be very simple, almost childishly so, and to lie well within the province of philosophical thought. Probably their very simplicity is the cause of error; we are not used to think about such simple abstract things, and a long training is necessary to secure even a partial immunity from error as soon as we diverge from the beaten track of thought.

The discovery of coordinate geometry, and also that of projective geometry about the same time, illustrate another fact which is being continually verified in the history of knowledge, namely, that some of the greatest discoveries are to be made among the most well-known topics. By the time that the seventeenth century had arrived, geometry had already been studied for over two thousand years, even if we date its rise with the Greeks. Euclid, taught in the University of Alexandria, being born about 330 B.C.; and he only systematized and extended the work of a long series of predecessors, some of them men of genius. After him generation after generation of mathematicians laboured at the improvement of the subject. Nor did the subject suffer from that fatal bar to progress, namely, that its study was confined to a narrow group of men of similar origin and outlook—quite the contrary was the case; by the seventeenth century it had passed through the minds of Egyptians and Greeks, of Arabs and of Germans. And yet, after all this labour devoted to it through so many ages by such diverse minds its most important secrets were yet to be discovered.

No one can have studied even the elements of elementary geometry without feeling the lack of some guiding method. Every proposition has to be proved by a fresh display of ingenuity; and a science for which this is true lacks the great requisite of scientific thought, namely, method. Now the especial point of coordinate geometry is that for the first time it introduced method. The remote deductions of a mathematical science are not of primary theoretical importance. The science has not been perfected, until it consists in essence of the exhibition of great allied methods by which information, on any desired topic which falls within its scope, can easily be obtained. The growth of a science is not primarily in bulk, but in ideas; and the more the ideas grow, the fewer are the deductions which it is worth while to write down. Unfortunately, mathematics is always encumbered by the repetition in text-books of numberless subsidiary propositions, whose importance has been lost by their absorption into the role of particular cases of more general truths—and, as we have already insisted, generality is the soul of mathematics.

Again, coordinate geometry illustrates another feature of mathematics which has already been pointed out, namely, that mathematical sciences as they develop dovetail into each other, and share the same ideas in common. It is not too much to say that the various branches of mathematics undergo a perpetual process of generalization, and that as they become generalized, they coalesce. Here again the reason springs from the very nature of the science, its generality, that is to say, from the fact that the science deals with the general truths which apply to all things in virtue of their very existence as things. In this connection the interest of coordinate geometry lies in the fact that it relates together geometry, which started as the science of space, and algebra, which has its origin in the science of number.

Let us now recall the main ideas of the two sciences, and

then see how they are related by Descartes' method of coordinates. Take algebra in the first place. We will not trouble ourselves about the imaginaries and will think merely of the real numbers with positive or negative signs. The fundamental idea is that of any number, the variable number, which is denoted by a letter and not by any definite numeral. We then proceed to the consideration of correlations between variables. For example, if x and y are two variables, we may conceive them as correlated by the equations $x + y = 1$, or by $x - y = 1$, or in any one of an indefinite number of other ways. This at once leads to the application of the idea of algebraic form. We think, in fact, of any correlation of some interesting type, thus rising from the initial conception of variable numbers to the secondary conception of variable correlations of numbers. Thus we generalize the correlation $x + y = 1$, into the correlation $ax + by = c$. Here a and b and c, being letters, stand for any numbers and are in fact themselves variables. But they are the variables which determine the variable correlation; and the correlation, when determined, correlates the variable numbers x and y. Variables, like a, b, and c above, which are used to determine the correlation are called "constants," or parameters. The use of the term "constant" in this connection for what is really a variable may seem at first sight to be odd; but it is really very natural. For the mathematical investigation is concerned with the relation between the variables x and y, after a, b, c are supposed to have been determined. So in a sense, relatively to x and y, the "constants" a, b, and c are constants. Thus $ax + by = c$ stands for the general example of a certain algebraic form, that is, for a variable correlation belonging to a certain class.

Again we generalize $x^2 + y^2 = 1$ into $ax^2 + by^2 = c$, or still further into $ax^2 + 2hxy + by^2 = c$, or, still further, into $ax^2 + 2hxy + by^2 + 2gx + 2fy = c$.

Here again we are led to variable correlations which are indicated by their various algebraic forms.

Now let us turn to geometry. The name of the science at once recalls to our minds the thought of figures and diagrams exhibiting triangles and rectangles and squares and circles, all in special relations to each other. The study of the simple properties of these figures is the subject matter of elementary geometry, as it is rightly presented to the beginner. Yet a moment's thought will show that this is not the true conception of the subject. It may be right for a child to commence his geometrical reasoning on shapes, like triangles and squares, which he has cut out with scissors. What, however, is a triangle? It is a figure marked out and bounded by three bits of three straight lines.

Now the boundary of spaces by bits of lines is a very complicated idea, and not at all one which gives any hope of exhibiting the simple general conceptions which should form the bones of the subject. We want something more simple and more general. It is this obsession with the wrong initial ideas—very natural and good ideas for the creation of first thoughts on the subject—which was the cause of the comparative sterility of the study of the science during so many centuries. Coordinate geometry, and Descartes its inventor, must have the credit of disclosing the true simple objects for geometrical thought.

In the place of a bit of a straight line, let us think of the whole of a straight line throughout its unending length in both directions. This is the sort of general idea from which to start our geometrical investigations. The Greeks never seem to have found any use for this conception which is now fundamental in all modern geometrical thought. Euclid always contemplates a straight line as drawn between two definite points, and is very careful to mention when it is to be produced beyond this segment. He never thinks of the line as an entity given once for all as a whole. This careful definition and limitation, so as to exclude an infinity not immediately apparent to the senses, was very characteristic of the Greeks in all their many activities. It is enshrined in the difference between Greek architecture and Gothic architecture, and between the Greek religion and the modern religion. The spire on a Gothic cathedral and the importance of the unbounded straight line in modern geometry are both emblematic of the transformation of the modern world.

The straight line, considered as a whole, is accordingly the root idea from which modern geometry starts. But then other sorts of lines occur to us, and we arrive at the conception of the complete curve which at every point of it exhibits some uniform characteristic, just as the straight line exhibits at all points the characteristic of straightness. For example, there is the circle which at all points exhibits the characteristic of being at a given distance from its centre, and again there is the ellipse, which is an oval curve, such that the sum of the two distances of any point on it from two fixed points, called its *foci*, is constant for all points on the curve. It is evident that a circle is merely a particular case of an ellipse when the two foci are superposed in the same point; for then the sum of the two distances is merely twice the radius of the circle. The ancients knew the properties of the ellipse and the circle and, of course, considered them as wholes. For example, Euclid never starts with mere segments (*i.e.*, bits) of circles, which are then prolonged. He always considers the whole circle as described. It is unfortunate that the circle is not the true fundamental line in geometry, so that his defective consideration of the straight line might have been of less consequence.

This general idea of a curve which at any point of it exhibits some uniform property is expressed in geometry by the term "locus." A locus is the curve (or surface, if we do not confine ourselves to a plane) formed by points, all of which possess some given property. To every property in relation to each other which points can have, there corresponds some locus, which consists of all the points possessing the property. In investigating the properties of a locus considered as a whole, we consider *any* point or points on the locus. Thus in geometry we again meet with the fundamental idea of the variable. Furthermore, in classifying loci under such headings as straight lines, circles, ellipses, etc., we again find the idea of form.

Accordingly, as in algebra we are concerned with variable numbers, correlations between variable numbers, and the classification of correlations into types by the idea of algebraic form; so in geometry we are concerned with variable points, variable points satisfying some condition so as form to a locus, and the classification of *loci* into types by the idea of conditions of the same form.

Now, the essence of coordinate geometry is the identification of the algebraic correlation with the geometrical locus. The point on a plane is represented in algebra by its two coordinates, x and y, and the condition satisfied by any point on the locus is represented by the corresponding correlation between x and y. Finally to correlations expressible in some general algebraic form, such as $ax+by=c$, there correspond loci of some general type, whose geometrical conditions are all of the same form. We have thus arrived at a position where we can effect a complete interchange in ideas and results between the two sciences. Each science throws light on the other, and itself gains immeasurably in power. It is impossible not to feel stirred at the thought of the emotions of men at certain historic moments of adventure and discovery—Columbus when he first saw the Western shore, Pizarro when he stared at the Pacific Ocean, Franklin when the electric spark came from the string of his kite, Galileo when he first turned his telescope to the heavens. Such moments are also granted to students in the abstract regions of thought, and high among them must be placed the morning when Descartes lay in bed and invented the method of coordinate geometry.

When one has once grasped the idea of coordinate geometry, the immediate question which starts to the mind is, What sort of loci correspond to the well-known algebraic forms? For example, the simplest among the general types of algebraic forms is $ax + by = c$. The sort of locus which corresponds to this is a straight line, and conversely to every straight line there corresponds an equation of this form. It is fortunate that the simplest among the geometrical loci should correspond to the simplest among the algebraic forms. Indeed, it is this general correspondence of geometrical and algebraic simplicity which gives to the whole subject its power. It springs from the fact that the connection between geometry and algebra is not casual and artificial, but deep-seated and essential. The equation which corresponds to a locus is called the equation "of" (or "to") the locus. Some examples of equations of straight lines will illustrate the subject.

Consider $y - x = 0$; here the a, b, and c, of the general form have been replaced by -1, 1, and 0 respectively. This line passes through the "origin," O, in the diagram and bisects the angle XOY. It is the line $L'OL$ of the diagram. The fact that it passes through the origin, O, is easily seen by observing that the equation is satisfied by putting $x = 0$ and $y = 0$ simultaneously, but 0 and 0 are the coordinates of O. In fact it is easy to generalize and to see by the same method that the equation of any line through the origin is of the form $ax + by = 0$. The locus of the equation $y + x = 0$

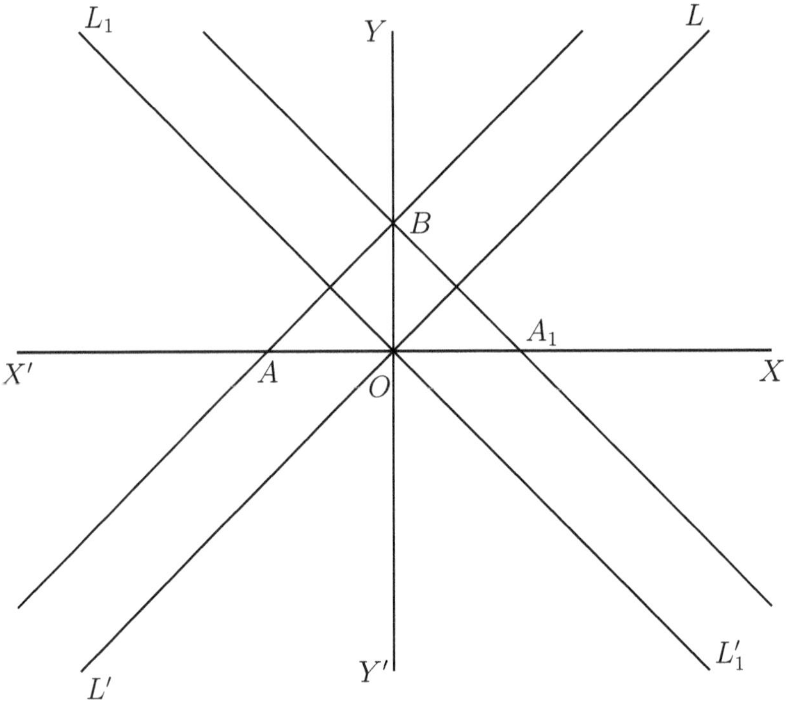

Fig. 14.

also passes through the origin and bisects the angle $X'OY$: it is the line L_1OL_1' of the diagram.

Consider $y - x = 1$: the corresponding locus does not pass through the origin. We therefore seek where it cuts the axes. It must cut the axis of x at some point of coordinates x and 0. But putting $y = 0$ in the equation, we get $x = -1$; so the coordinates of this point (A) are 1 and 0. Similarly the point (B) where the line cuts the axis OY are 0 and 1.

The locus is the line AB in the figure and is parallel to LOL'. Similarly $y + x = 1$ is the equation of line A_1B of the figure; and the locus is parallel to $L_1OL'_1$. It is easy to prove the general theorem that two lines represented by equations of the forms $ax + by = 0$ and $ax + by = c$ are parallel.

The group of loci which we next come upon are sufficiently important to deserve a chapter to themselves. But before going on to them we will dwell a little longer on the main ideas of the subject.

The position of any point P is determined by arbitrarily choosing an origin, O, two axes, OX and OY, at right-angles, and then by noting its coordinates x and y, *i.e.* OM and PM (*cf.* fig. 13). Also, as we have seen in the last chapter, P can be determined by the "vector" OP, where the idea of the vector includes a determinate direction as well as a determinate length. From an abstract mathematical point of view the idea of an arbitrary origin may appear artificial and clumsy, and similarly for the arbitrarily drawn axes, OX and OY. But in relation to the application of mathematics to the event of the Universe we are here symbolizing with direct simplicity the most fundamental fact respecting the outlook on the world afforded to us by our senses. We each of us refer our sensible perceptions of things to an origin which we call "here": our location in a particular part of space round which we group the whole Universe is the essential fact of our bodily existence. We can imagine beings who observe all phenomena in all space with an equal eye, unbiassed in favour of any part. With us it is otherwise, a cat at our feet claims more attention than an earthquake at Cape Horn, or than the destruction of a world in the Milky Way. It is true that in making a common stock of our knowledge with our fellowmen, we have to waive something of the strict egoism of our own individual "here." We substitute "nearly here" for "here"; thus we measure miles from the town hall of the nearest town, or from the capital of the country. In measuring the earth, men of science will put the origin at the earth's centre; astronomers even rise to the extreme altruism of putting their origin inside the sun. But, far as this last origin may be, and even if we go further to some convenient point amid the nearer fixed stars, yet, compared to the immeasurable infinities of space, it remains true that our first procedure in exploring the Universe is to fix upon an origin "nearly here."

Again the relation of the coordinates OM and MP (*i.e.* x and y) to the vector OP is an instance of the famous parallelogram law, as can easily be seen (*cf.* fig. 8) by completing the parallelogram $OMPN$. The idea of the "vector" OP, that is, of a directed magnitude, is the root-idea of physical science. Any moving body has a certain magnitude of velocity in a certain direction, that is to say, its velocity is a directed magnitude, a vector. Again a force has a certain magnitude and has a definite direction. Thus, when in analytical geometry the ideas of the "origin," of "coordinates," and of "vectors" are introduced, we are studying the abstract conceptions which correspond to the fundamental facts of the physical world.

CHAPTER X

CONIC SECTIONS

WHEN the Greek geometers had exhausted, as they thought, the more obvious and interesting properties of figures made up of straight lines and circles, they turned to the study of other curves; and, with their almost infallible instinct for hitting upon things worth thinking about, they chiefly devoted themselves to conic sections, that is, to the curves in which planes would cut the surfaces of circular cones. The man who must have the credit of inventing the study is Menaechmus (born 375 B.C. and died 325 B.C.); he was a pupil of Plato and one of the tutors of Alexander the Great. Alexander, by the by, is a conspicuous example of the advantages of good tuition, for another of his tutors was the philosopher Aristotle. We may suspect that Alexander found Menaechmus rather a dull teacher, for it is related that he asked for the proofs to be made shorter. It was to this request that Menaechmus replied: "In the country there are private and even royal roads, but in geometry there is only one road for all." This reply no doubt was true enough in the sense in which it would have been immediately understood by Alexander. But if Menaechmus thought that his proofs could not be shortened, he was grievously mistaken; and most modern mathematicians would be horribly bored, if they were compelled to study the Greek proofs of the properties of conic sections. Nothing illustrates better the gain in power which is obtained by the introduction of relevant ideas into a science than to observe the progressive shortening of proofs which accompanies the growth of richness in idea. There is a certain type of mathematician who is always rather impatient at delaying over the ideas of a subject: he is anxious at once to get on to the proofs of "important" problems. The history of the science is entirely against him. There are royal roads in science; but those who first tread them are men of genius and not kings.

The way in which conic sections first presented themselves to mathematicians was as follows: think of a cone (*cf.* fig. 15), whose vertex (or point) is V, standing on a circular base STU. For example, a conical shade to an electric light is often an example of such a surface. Now let the "generating" lines which pass through V and lie on the surface be all produced backwards; the result is a double cone, and PQR is another circular cross section on the opposite side of V to the cross section STU. The axis of the cone CVC' passes through all the centres of these circles and is perpendicular to their planes, which are parallel to each other. In the diagram the parts of the curves which are supposed to lie behind the plane of the paper are dotted lines, and the parts on the plane or in front of it are continuous lines. Now suppose this double cone is cut by a plane not perpendicular to the axis CVC', or at least not necessarily perpendicular to it. Then three cases can arise:—

(1) The plane may cut the cone in a closed oval curve, such as $ABA'B'$ which lies entirely on one of the two half-

cones. In this case the plane will not meet the other half-cone at all. Such a curve is called an ellipse; it is an oval curve. A particular case of such a section of the cone is when the plane is perpendicular to the axis CVC', then the section, such as STU or PQR, is a circle. Hence a circle is a particular case of the ellipse.

(2) The plane may be parallelled to a tangent plane touching the cone along one of its "generating" lines as for example the plane of the curve $D_1A_1D_1'$ in the diagram is parallel to the tangent plane touching the cone along the generating line VS; the curve is still confined to one of the half-cones, but it is now not a closed oval curve, it goes on endlessly as long as the generating lines of the half-cone are produced away from the vertex. Such a conic section is called parabola.

(3) The plane may cut both the half-cones, so that the complete curve consists of two detached portions, or "branches" as they are called, this case is illustrated by the two branches $G_2A_2G_2'$ and $L_2A_2'L_2'$ which together make up the curve. Neither branch is closed, each of them spreading out endlessly as the two half-cones are prolonged away from the vertex. Such a conic section is called a hyperbola.

There are accordingly three types of conic sections, namely, ellipses, parabolas, and hyperbolas. It is easy to see that, in a sense, parabolas are limiting cases lying between ellipses and hyperbolas. They form a more special sort and have to satisfy a more particular condition. These three names are apparently due to Apollonius of Perga (born about 260 B.C., and died about 200 B.C.), who wrote a systematic treatise on conic sections which remained the standard work till the sixteenth century.

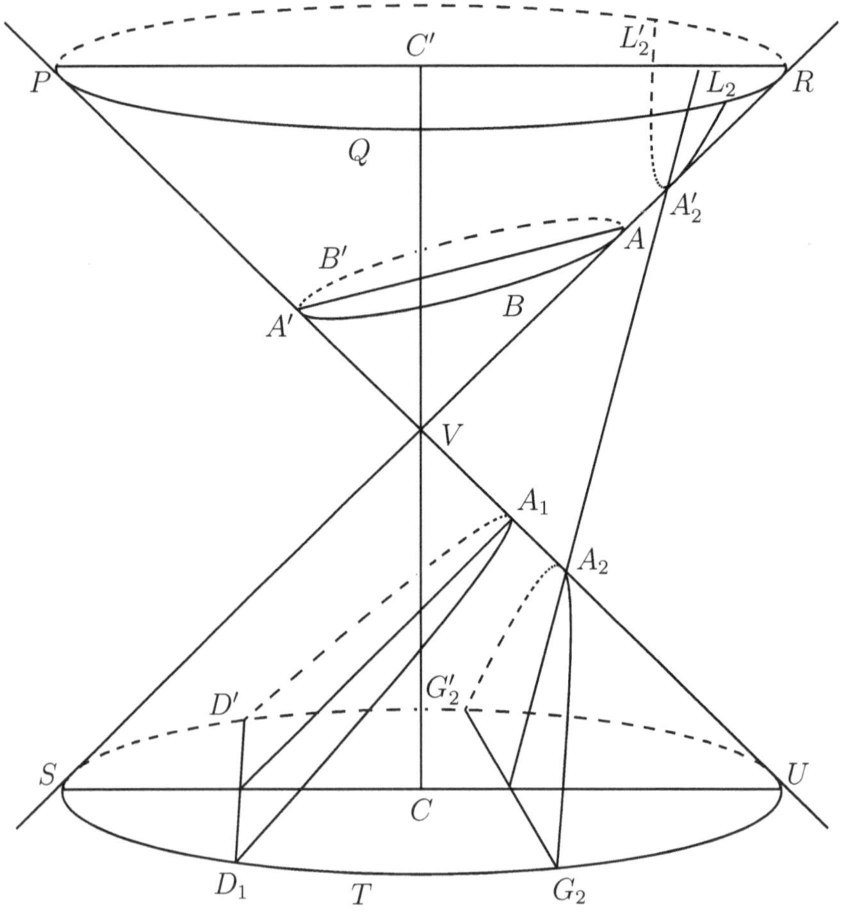

Fig. 15.

It must at once be apparent how awkward and difficult the investigation of the properties of these curves must have been to the Greek geometers. The curves are plane curves, and yet their investigation involves the drawing in perspective of a solid figure. Thus in the diagram given above we have practically drawn no subsidiary lines and yet the figure is sufficiently complicated. The curves are plane curves, and it seems obvious that we should be able to define them without going beyond the plane into a solid figure. At the

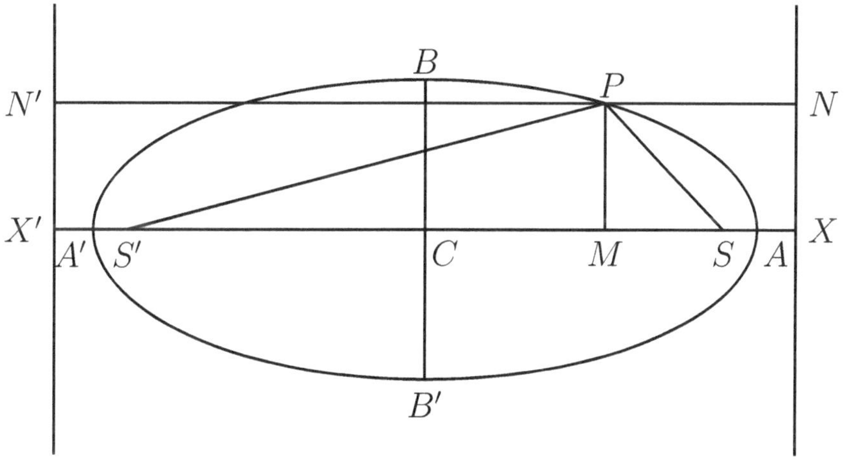

Fig. 16.

same time, just as in the "solid" definition there is one uniform method of definition—namely, the section of a cone by a plane—which yields three cases, so in any "plane" definition there also should be one uniform method of procedure which falls into three cases. Their shapes when drawn on

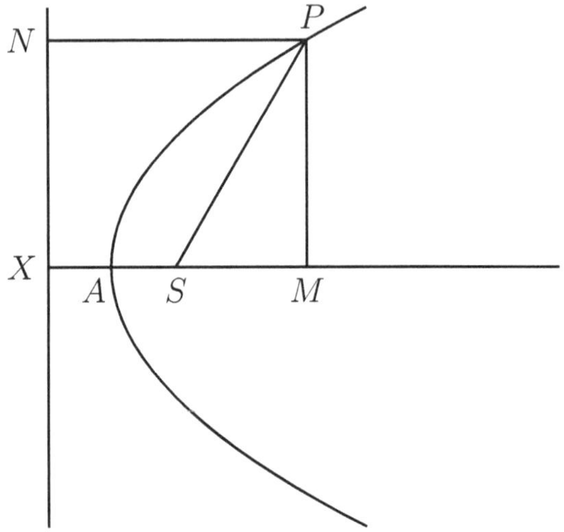

Fig. 17.

their planes are those of the curved lines in the three figures 16, 17, and 18. The points A and A' in the figures are called the vertices and the line AA' the major axis. It will be noted that a parabola (*cf.* fig. 17) has only one vertex. Apollonius proved* that the ratio of PM^2 to $AM \cdot MA'$ $\left(i.e.\ \dfrac{PM^2}{AM \cdot MA'}\right)$ remains constant both for the ellipse and the hyperbola (figs. 16 and 18), and that the ratio of PM^2 to AM is constant for the parabola of fig. 17; and he bases most of his work on this fact. We are evidently advancing towards the desired uniform definition which does not go out of the plane; but have not yet quite attained to uniformity.

* *Cf.* Ball, *loc. cit.*, for this account of Apollonius and Pappus.

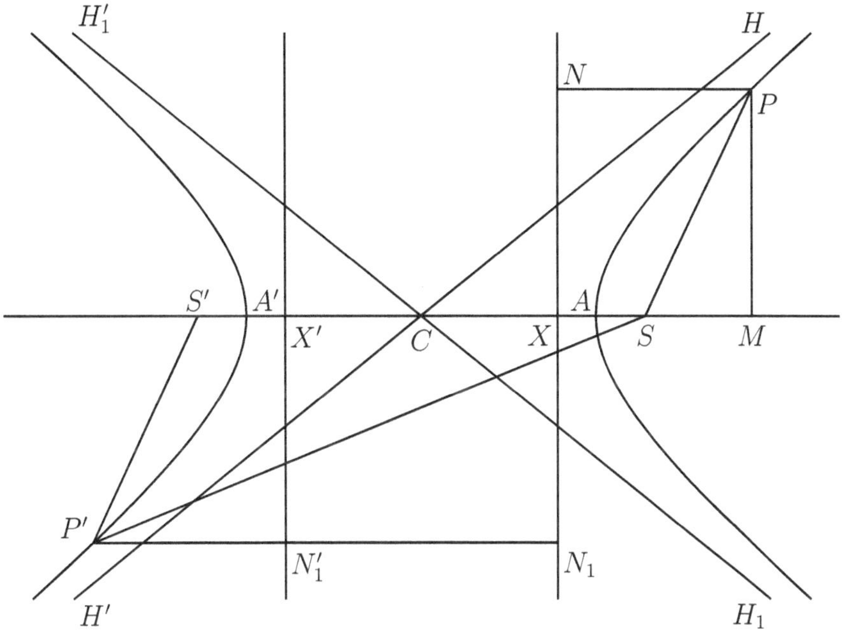

Fig. 18.

In the diagrams 16 and 18, two points, S and S', will be seen marked, and in diagram 17 one point, S. These are the *foci* of the curves, and are points of the greatest importance. Apollonius knew that for an ellipse the sum of SP and $S'P$ (*i.e.* $SP + S'P$) is constant as P moves on the curve, and is equal to AA'. Similarly for a hyperbola the difference $S'P - SP$ is constant, and equal to AA' when P is on one branch, and the difference $SP' - S'P'$ is constant and equal to AA' when P' is on the other branch. But no corresponding point seemed to exist for the parabola.

Finally 500 years later the last great Greek geometer, Pappus of Alexandria, discovered the final secret which completed this line of thought. In the diagrams 16 and 18 will be seen two lines, XN and $X'N'$, and in diagram 17 the single line, XN. These are the directrices of the curves, two each for the ellipse and the hyperbola, and one for the parabola. Each directrix corresponds to its nearer focus. The characteristic property of a focus, S, and its corresponding directrix, XN, for any one of the three types of curve, is that the ratio SP to PN $\left(i.e.\ \dfrac{SP}{PN}\right)$ is constant, where PN is the perpendicular on the directrix from P, and P is any point on the curve. Here we have finally found the desired property of the curves which does not require us to leave the plane, and is stated uniformly for all three curves. For ellipses the ratio* is less than 1, for parabolas it is equal to 1, and for hyperbolas it is greater than 1.

When Pappus had finished his investigations, he must have felt that, apart from minor extensions, the subject was practically exhausted; and if he could have foreseen the history of science for more than a thousand years, it would have confirmed his belief. Yet in truth the really fruitful ideas in connection with this branch of mathematics had not yet been even touched on, and no one had guessed their supremely important applications in nature. No more impressive warning can be given to those who would confine knowledge and re-

* *Cf.* Note B, p. 207.

search to what is apparently useful, than the reflection that conic sections were studied for eighteen hundred years merely as an abstract science, without a thought of any utility other than to satisfy the craving for knowledge on the part of mathematicians, and that then at the end of this long period of abstract study, they were found to be the necessary key with which to attain the knowledge of one of the most important laws of nature.

Meanwhile the entirely distinct study of astronomy had been going forward. The great Greek astronomer Ptolemy (died 168 A.D.) published his standard treatise on the subject in the University of Alexandria, explaining the apparent motions among the fixed stars of the sun and planets by the conception of the earth at rest and the sun and the planets circling round it. During the next thirteen hundred years the number and the accuracy of the astronomical observations increased, with the result that the description of the motions of the planets on Ptolemy's hypothesis had to be made more and more complicated. Copernicus (born 1473 A.D. and died 1543 A.D.) pointed out that the motions of these heavenly bodies could be explained in a simpler manner if the sun were supposed to rest, and the earth and planets were conceived as moving round it. However, he still thought of these motions as essentially circular, though modified by a set of small corrections arbitrarily superimposed on the primary circular motions. So the matter stood when Kepler was born at Stuttgart in Germany in 1571 A.D. There were two sciences, that of the geometry of conic sections and that of astronomy, both of which had been studied from a remote antiquity without a suspicion of any connection between the two. Kepler was an astronomer, but he was also an able geometer, and on the subject of conic sections had arrived at ideas in advance of his time. He is only one of many examples of the falsity of the idea that success in scientific research demands an exclusive absorption in one narrow line of study. Novel ideas are more apt to spring from an unusual assortment of knowledge—not necessarily from vast knowledge, but from a thorough conception of the methods and ideas of distinct lines of thought. It will be remembered that Charles Darwin was helped to arrive at his conception of the law of evolution by reading Malthus' famous *Essay on Population*, a work dealing with a different subject—at least, as it was then thought.

Kepler enunciated three laws of planetary motion, the first two in 1609, and the third ten years later. They are as follows:

(1) The orbits of the planets are ellipses, the sun being in the focus.

(2) As a planet moves in its orbit, the radius vector from the sun to the planet sweeps out equal areas in equal times.

(3) The squares of the periodic times of the several planets are proportional to the cubes of their major axes.

These laws proved to be only a stage towards a more fundamental development of ideas. Newton (born 1642 A.D. and died 1727 A.D.) conceived the idea of universal gravitation, namely, that any two pieces of matter attract each other with a force proportional to the product of their masses and inversely proportional to the square of their distance from each other. This sweeping general law, coupled with the three laws of motion which he put into their final general shape, proved adequate to explain all astronomical phenomena, including Kepler's laws, and has formed the basis of modern physics. Among other things he proved that comets might move in very elongated ellipses, or in parabolas, or in hyperbolas, which are nearly parabolas. The comets which return—such as Halley's comet—must, of course, move in ellipses. But the essential step in the proof of the law of gravitation, and even in the suggestion of its initial conception, was the verification of Kepler's laws connecting the motions of the planets with the theory of conic sections.

From the seventeenth century onwards the abstract theory of the curves has shared in the double renaissance of geometry due to the introduction of coordinate geometry and of projective geometry. In projective geometry the fundamental ideas cluster round the consideration of sets (or pencils, as they are called) of lines passing through a common point (the vertex of the "pencil"). Now (*cf.* fig. 19) if A, B, C, D, be any four fixed points on a conic section and P be a variable point on the curve, the pencil of lines PA, PB, PC, and PD, has a special property, known as the constancy of its cross ratio. It will suffice here to say that cross ratio is a fundamental idea in projective geometry. For projective geometry this is really the definition of the curves, or some analogous property which is really equivalent to it. It will be

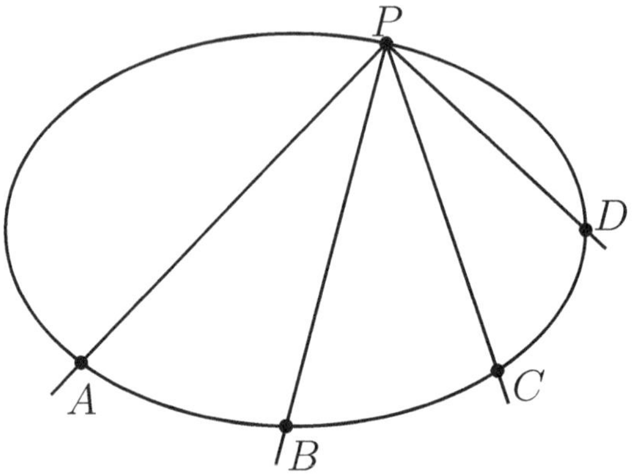

Fig. 19.

seen how far in the course of ages of study we have drifted away from the old original idea of the sections of a circular cone. We know now that the Greeks had got hold of a minor property of comparatively slight importance; though by some divine good fortune the curves themselves deserved all the attention which was paid to them. This unimportance of the "section" idea is now marked in ordinary mathematical phraseology by dropping the word from their names. As often as not, they are now named merely "conics" instead of "conic sections."

Finally, we come back to the point at which we left coordinate geometry in the last chapter. We had asked what was the type of *loci* corresponding to the general algebraic form $ax + by = c$, and had found that it was the class of straight lines in the plane. We had seen that every straight line possesses an equation of this form, and that every equation of this form corresponds to a straight line. We now wish to go on to the next general type of algebraic forms. This is evidently to be obtained by introducing terms involving x^2 and xy and y^2. Thus the new general form must be written:—

$$ax^2 + 2hxy + by^2 + 2gx + 2fy + c = 0.$$

What does this represent? The answer is that (when it represents any locus) it always represents a conic section, and, furthermore, that the equation of every conic section can always be put into this shape. The discrimination of the particular sorts of conics as given by this form of equation is very easy. It entirely depends upon the consideration of $ab - h^2$, where a, b, and h, are the "constants" as written above. If $ab - h^2$ is a positive number, the curve is an ellipse; if $ab - h^2 = 0$, the curve is a parabola: and if $ab - h^2$ is a negative number, the curve is a hyperbola.

For example, put $a = b = 1$, $h = g = f = 0$, $c = -4$. We then get the equation $x^2 + y^2 - 4 = 0$. It is easy to prove that this is the equation of a circle, whose centre is at the origin, and radius is 2 units of length. Now $ab - h^2$ becomes $1 \times 1 - 0^2$, that is, 1, and is therefore positive. Hence the circle is a particular case of an ellipse, as it ought to be. Generalising, the equation of any circle can be put into the form $a(x^2 + y^2) + 2gx + 2fy + c = 0$. Hence $ab - h^2$ becomes $a^2 - 0$, that is, a^2, which is necessarily positive. Accordingly all circles satisfy the condition for ellipses. The general form of the equation of a parabola is

$$(dx + ey)^2 + 2gx + 2fy + c = 0,$$

so that the terms of the second degree, as they are called, can be written as a perfect square. For squaring out, we get

$$d^2x^2 + 2dexy + e^2y^2 + 2gx + 2fy + c;$$

so that by comparison $a = d^2$, $h = de$, $b = e^2$, and therefore $ab - h^2 = d^2e^2 - (de)^2 = 0$. Hence the necessary condition is automatically satisfied. The equation $2xy - 4 = 0$, where $a = b = g = f = 0$, $h = 1$, $c = -4$, represents a hyperbola. For the condition $ab - h^2$ becomes $0 - 1^2$, that is, -1, which is negative.

The limitation, introduced by saying that, *when the general equation represents any locus*, it represents a conic section, is necessary, because some particular cases of the general equation represent no real locus. For example $x^2 + y^2 + 1 = 0$ can be satisfied by no real values of x and y. It is usual to say that the locus is now one composed of imaginary points. But this idea of imaginary points in geometry is really one of great complexity, which we will not now enter into.

Some exceptional cases are included in the general form of the equation which may not be immediately recognized as conic sections. By properly choosing the constants the equation can be made to represent two straight lines. Now two intersecting straight lines may fairly be said to come under the Greek idea of a conic section. For, by referring to the picture of the double cone above, it will be seen that some planes through the vertex, V, will cut the cone in a pair of straight lines intersecting at V. The case of two parallel straight lines can be included by considering a circular cylinder as a particular case of a cone. Then a plane, which cuts it and is parallel to its axis, will cut it in two parallel straight lines. Anyhow, whether or no the ancient Greek would have allowed these special cases to be called conic sections, they are certainly included among the curves represented by the general algebraic form of the second degree. This fact is worth noting; for it is characteristic of modern mathematics to include among general forms all sorts of particular cases which would formerly have received special treatment. This is due to its pursuit of generality.

CHAPTER XI

FUNCTIONS

THE mathematical use of the term function has been adopted also in common life. For example, "His temper is a function of his digestion," uses the term exactly in this mathematical sense. It means that a rule can be assigned which will tell you what his temper will be when you know how his digestion is working. Thus the idea of a "function" is simple enough, we only have to see how it is applied in mathematics to variable numbers. Let us think first of some concrete examples: If a train has been travelling at the rate of twenty miles per hour, the distance (s miles) gone after any number of hours, say t, is given by $s = 20 \times t$; and s is called a function of t. Also $20 \times t$ is the function of t with which s is identical. If John is one year older than Thomas, then, when Thomas is at any age of x years, John's age (y years) is given by $y = x + 1$; and y is a function of x, namely, is the function $x + 1$.

In these examples t and x are called the "arguments" of the functions in which they appear. Thus t is the argument of the function $20 \times t$, and x is the argument of the function $x + 1$. If $s = 20 \times t$, and $y = x + 1$, then s and y are called the "values" of the functions $20 \times t$ and $x + 1$ respectively.

Coming now to the general case, we can define a function in mathematics as a correlation between two variable numbers, called respectively the argument and the value of the function, such that whatever value be assigned to the "argument of the function" the "value of the function" is definitely (*i.e.* uniquely) determined. The converse is not necessarily true, namely, that when the value of the function is determined the argument is also uniquely determined. Other functions of the argument x are $y = x^2$, $y = 2x^2 + 3x + 1$, $y = x$, $y = \log x$, $y = \sin x$. The last two functions of this group will be readily recognizable by those who understand a little algebra and trigonometry. It is not worth while to delay now for their explanation, as they are merely quoted for the sake of example.

Up to this point, though we have defined what we mean by a function in general, we have only mentioned a series of special functions. But mathematics, true to its general methods of procedure, symbolizes the general idea of any function. It does this by writing $F(x)$, $f(x)$, $g(x)$, $\phi(x)$, etc., for any function of x, where the argument x is placed in a bracket, and some letter like F, f, g, ϕ, etc., is prefixed to the bracket to stand for the function. This notation has its defects. Thus it obviously clashes with the convention that the single letters are to represent variable numbers; since here F, f, g, ϕ, etc., prefixed to a bracket stand for variable functions. It would be easy to give examples in which we can only trust to common sense and the context to see what is meant. One way of evading the confusion is by using Greek letters (*e.g.* ϕ as above) for functions; another way is to keep to f and F (the initial letter of function) for the functional letter, and, if other variable functions have to be symbolized, to take an adjacent letter like g.

With these explanations and cautions, we write $y = f(x)$, to denote that y is the value of some undetermined function of the argument x; where $f(x)$ may stand for anything such as $x+1$, $x^2 - 2x + 1$, $\sin x$, $\log x$, or merely for x itself. The essential point is that when x is given, then y is thereby definitely determined. It is important to be quite clear as to the generality of this idea. Thus in $y = f(x)$, we may determine, if we choose, $f(x)$ to mean that when x is an integer, $f(x)$ is zero, and when x has any other value, $f(x)$ is 1. Accordingly, putting $y = f(x)$, with this choice for the meaning of f, y is either 0 or 1 according as the value of x is integral or otherwise. Thus $f(1) = 0$, $f(2) = 0$, $f(\frac{2}{3}) = 1$, $f(\sqrt{2}) = 1$, and so on. This choice for the meaning of $f(x)$ gives a perfectly good function of the argument x according to the general definition of a function.

A function, which after all is only a sort of correlation between two variables, is represented like other correlations by a graph, that is in effect by the methods of coordinate geometry. For example, fig. 2 in Chapter II. is the graph of the function $\frac{1}{v}$ where v is the argument and p the value of the function. In this case the graph is only drawn for positive values of v, which are the only values possessing any meaning for the physical application considered in that chapter. Again in fig. 14 of Chapter IX. the whole length of the line AB, unlimited in both directions, is the graph of the function $x+1$, where x is the argument and y is the value of the function; and in the same figure the unlimited line A_1B is the graph of the function $1 - x$, and the line LOL' is the graph of the function x, x being the argument and y the value of the function.

These functions, which are expressed by simple algebraic formulæ, are adapted for representation by graphs. But for some functions this representation would be very misleading without a detailed explanation, or might even be impossible. Thus, consider the function mentioned above, which has the value 1 for all values of its argument x, except those which are integral, *e.g.* except for $x = 0$, $x = 1$, $x = 2$, etc., when it has the value 0. Its appearance on a graph would be that of the straight line ABA' drawn parallel to the axis XOX' at

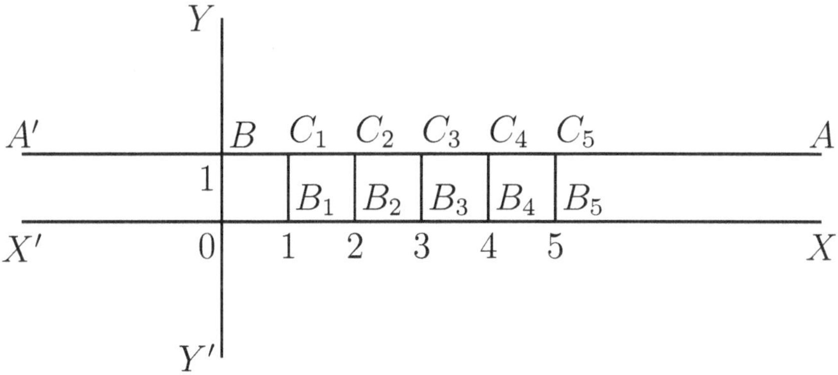

Fig. 20.

a distance from it of 1 unit of length. But the points, B, C_1, C_2, C_3, C_4, etc., corresponding to the values 0, 1, 2, 3, 4, etc., of the argument x, are to be omitted, and instead of them the points O, B_1, B_2, B_3, B_4, etc., on the axis OX, are to be taken. It is easy to find functions for which the graphical representation is not only inconvenient but impossible. Functions which do not lend themselves to graphs are important in the higher mathematics, but we need not concern ourselves further about them here.

The most important division between functions is that between continuous and discontinuous functions. A function is continuous when its value only alters gradually for gradual alterations of the argument, and is discontinuous when it can alter its value by sudden jumps. Thus the two functions $x+1$ and $1-x$, whose graphs are depicted as straight lines in fig. 14 of Chapter IX., are continuous functions, and so is the function $\frac{1}{v}$, depicted in Chapter II., if we only think of positive values of v. But the function depicted in fig. 20 of this chapter is discontinuous since at the values $x = 1$, $x = 2$, etc., of its argument, its value gives sudden jumps.

Let us think of some examples of functions presented to us in nature, so as to get into our heads the real bearing of continuity and discontinuity. Consider a train in its journey along a railway line, say from Euston Station, the terminus in London of the London and North-Western Railway. Along the line in order lie the stations of Bletchley and Rugby. Let t be the number of hours which the train has been on its journey from Euston, and s be the number of miles passed over. Then s is a function of t, *i.e.* is the variable value corresponding to the variable argument t. If we know the circumstances of the train's run, we know s as soon as any special value of t is given. Now, miracles apart, we may confidently assume that s is a continuous function of t. It is impossible to allow for the contingency that we can trace the train continuously from Euston to Bletchley, and that then, without any intervening time, however short, it should appear at Rugby. The idea is too fantastic to enter into our calculation: it contemplates possibilities not to be found outside the *Arabian Nights*; and even in those tales sheer discontinuity of motion hardly enters into the imagination, they do not dare to tax our credulity with anything more than very unusual speed. But unusual speed is no contradiction to the great law of continuity of motion which appears to hold in nature. Thus light moves at the rate of about $190,000$ miles per second and comes to us from the sun in seven or eight minutes; but, in spite of this speed, its distance travelled is always a continuous function of the time.

It is not quite so obvious to us that the velocity of a body is invariably a continuous function of the time. Consider the train at any time t: it is moving with some definite velocity, say v miles per hour, where v is zero when the train is at rest in a station and is negative when the train is backing. Now we readily allow that v cannot change its value suddenly for a big, heavy train. The train certainly cannot be running at forty miles per hour from 11.45 a.m. up to noon, and then suddenly, without any lapse of time, commence running at 50 miles per hour. We at once admit that the change of velocity will be a gradual process. But how about sudden blows of adequate magnitude? Suppose two trains collide; or, to take smaller objects, suppose a man kicks a football. It certainly appears to our sense as though the football began suddenly to move. Thus, in the case of velocity our senses do not revolt at the idea of its being a discontinuous function of the time, as they did at the idea of the train being instantaneously transported from Bletchley to Rugby. As a matter of fact, if the laws of motion, with their conception of mass, are true, there is no such thing as discontinuous velocity in nature. Anything that appears to our senses as discontinuous change of velocity must, according to them, be considered to be a case of gradual change which is too quick to be perceptible to us. It would be rash, however, to rush into the generalization that no discontinuous functions are presented to us in nature. A man who, trusting that the mean height of the land above sea-level between London and Paris was a continuous function of the distance from London, walked at night on Shakespeare's Cliff by Dover in contemplation of the Milky Way, would be dead before he had had time to rearrange his ideas as to the necessity of caution in scientific conclusions.

It is very easy to find a discontinuous function, even if we confine ourselves to the simplest of the algebraic formulæ. For example, take the function $y = \frac{1}{x}$, which we have already considered in the form $p = \frac{1}{v}$, where v was confined to positive values. But now let x have any value, positive or

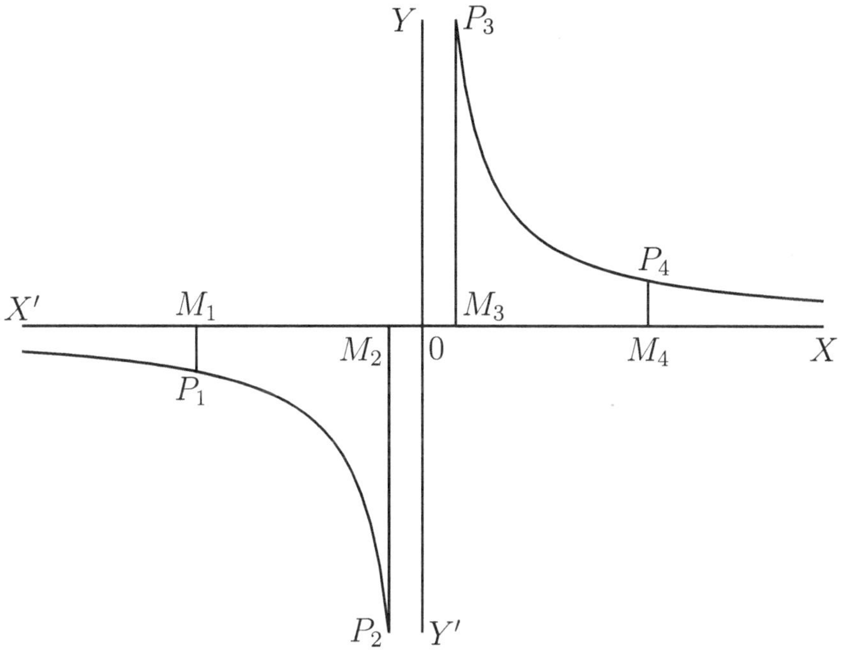

Fig. 21.

negative. The graph of the function is exhibited in fig. 21. Suppose x to change continuously from a large negative value through a numerically decreasing set of negative values up to 0, and thence through the series of increasing positive values. Accordingly, if a moving point, M, represents x on XOX', M starts at the extreme left of the axis XOX' and successively moves through M_1, M_2, M_3, M_4, etc. The corresponding points on the function are P_1, P_2, P_3, P_4, etc. It is easy to see that there is a point of discontinuity at $x = 0$, *i.e.* at the origin O. For the value of the function on the negative (left) side of the origin becomes endlessly great, but negative, and the function reappears on the positive (right) side as endlessly great but positive. Hence, however small we take the length M_2M_3, there is a finite jump between the values of the function at M_2 and M_3. Indeed, this case has the peculiarity that the smaller we take the length between M_2 and M_3, so long as they enclose the origin, the bigger is the jump in value of the function between them. This graph brings out, what is also apparent in fig. 20 of this chapter, that for many functions the discontinuities only occur at isolated points, so that by restricting the values of the argument we obtain a continuous function for these remaining values. Thus it is evident from fig. 21 that in $y = \dfrac{1}{x}$, if we keep to positive values only and exclude the origin, we obtain a continuous function. Similarly the same function, if we keep to negative values only, excluding the origin, is continuous. Again the function which is graphed in fig. 20 is continuous between B and C_1, and between C_1 and C_2, and between C_2 and C_3, and so on, always in each case excluding the end points. It is, however, easy to find functions such that their discontinuities occur at all points. For example, consider a function $f(x)$, such that when x is any fractional number $f(x) = 1$, and when x is any incommensurable number $f(x) = 2$. This function is discontinuous at all points.

Finally, we will look a little more closely at the definition of continuity given above. We have said that a function is continuous when its value only alters gradually for gradual alterations of the argument, and is discontinuous when it can alter its value by sudden jumps. This is exactly the sort of definition which satisfied our mathematical forefathers and no longer satisfies modern mathematicians. It is worth while to spend some time over it; for when we understand the modern objections to it, we shall have gone a long way towards the understanding of the spirit of modern mathematics. The whole difference between the older and the newer mathematics lies in the fact that vague half-metaphorical terms like "gradually" are no longer tolerated in its exact statements. Modern mathematics will only admit statements and definitions and arguments which exclusively employ the few simple ideas about number and magnitude and variables on which the science is founded. Of two numbers one can be greater or less than the other; and one can be such and such a multiple of the other; but there is no relation of "graduality" between two numbers, and hence the term is inadmissible. Now this may seem at first sight to be great pedantry. To this charge there are two answers. In the first place, during the first half of the nineteenth century it was found by some great mathematicians, especially Abel in Sweden, and Weierstrass in Germany, that large parts of mathematics as enunciated in the old happy-go-lucky manner were simply wrong. Macaulay in his essay on Bacon contrasts the certainty of mathematics with the uncertainty of philosophy; and by way of a rhetorical example he says, "There has been no reaction against Taylor's theorem." He could not have chosen a worse example. For, without having made an examination of English text-books on mathematics contemporary with the publication of this essay, the assumption is a fairly safe one that Taylor's theorem was enunciated and proved wrongly in every one of them. Accordingly, the anxious precision of modern mathematics is necessary for accuracy. In the second place it is necessary for research. It makes for clearness of thought, and thence for boldness of thought and for fertility in trying new combinations of ideas. When the initial statements are vague and slipshod, at every subsequent stage of thought common sense has to step in to limit applications and to explain meanings. Now in creative thought common sense is a bad master. Its sole criterion for judgment is that the new ideas shall look like the old ones. In other words it can only act by suppressing originality.

In working our way towards the precise definition of continuity (as applied to functions) let us consider more closely the statement that there is no relation of "graduality" between numbers. It may be asked, Cannot one number be only slightly greater than another number, or in other words, cannot the difference between the two numbers be small? The whole point is that in the abstract, apart from some arbitrarily assumed application, there is no such thing as a great or a small number. A million miles is a small number of miles for an astronomer investigating the fixed stars, but a million pounds is a large yearly income. Again, one-quarter is a large fraction of one's income to give away in charity, but is a small fraction of it to retain for private use. Examples can be accumulated indefinitely to show that great or small in any absolute sense have no abstract application to numbers. We can say of two numbers that one is greater or smaller than another, but not without specification of particular circumstances that any one number is great or small. Our task therefore is to define continuity without any mention of a "small" or "gradual" change in value of the function.

In order to do this we will give names to some ideas, which will also be useful when we come to consider limits and the differential calculus.

An "interval" of values of the argument x of a function $f(x)$ is all the values lying between some two values of the argument. For example, the interval between $x = 1$ and $x = 2$ consists of all the values which x can take lying between 1 and 2, *i.e.* it consists of all the real numbers between 1 and 2. But the bounding numbers of an interval need not be integers. An interval of values of the argument *contains* a number a, when a is a member of the interval. For example, the interval between 1 and 2 contains $\frac{3}{2}$, $\frac{5}{3}$, $\frac{7}{4}$, and so on.

A set of numbers approximates to a number a within a *standard* k, when the numerical difference between a and every number of the set is less than k. Here k is the "standard of approximation." Thus the set of numbers 3, 4, 6, 8, approximates to the number 5 within the standard 4. In this case the standard 4 is not the smallest which could have been chosen, the set also approximates to 5 within any of the standards 3.1 or 3.01 or 3.001. Again, the numbers, 3.1,

3.141, 3.1415, 3.14159 approximate to 3.13102 within the standard .032, and also within the smaller standard .03103.

These two ideas of an interval and of approximation to a number within a standard are easy enough; their only difficulty is that they look rather trivial. But when combined with the next idea, that of the "neighbourhood" of a number, they form the foundation of modern mathematical reasoning. What do we mean by saying that something is true for a function $f(x)$ in the neighbourhood of the value a of the argument x? It is this fundamental notion which we have now got to make precise.

The values of a function $f(x)$ are said to possess a characteristic in the "neighbourhood of a" when some interval can be found, which (i) contains the number a not as an end-point, and (ii) is such that every value of the function for arguments, other than a, lying within that interval possesses the characteristic. The value $f(a)$ of the function for the argument a may or may not possess the characteristic. Nothing is decided on this point by statements about the *neighbourhood* of a.

For example, suppose we take the particular function x^2. Now *in the neighbourhood of* 2, the values of x^2 are less than 5. For we can find an interval, *e.g.* from 1 to 2.1, which (i) contains 2 not as an end-point, and (ii) is such that, for values of x lying within it, x^2 is less than 5.

Now, combining the preceding ideas we know what is meant by saying that *in the neighbourhood of* a the function $f(x)$ approximates to c within the *standard* k. It means that some interval can be found which (i) includes a not as an end-point, and (ii) is such that all values of $f(x)$, where x lies in the interval and is not a, differ from c by less than k. For example, in the neighbourhood of 2, the function $\sqrt{x}$ approximates to 1.41425 within the standard .0001. This is true because the square root of 1.99996164 is 1.4142 and the square root of 2.00024449 is 1.4143; hence for values of x lying in the interval 1.99996164 to 2.00024449, which contains 2 not as an end-point, the values of the function $\sqrt{x}$ all lie between 1.4142 and 1.4143, and they therefore all differ from 1.41425 by less than .0001. In this case we can, if we like, fix a smaller standard of approximation, namely .000051 or .0000501. Again, to take another example, in the neighbourhood of 2 the function x^2 approximates to 4 within the standard .5. For $(1.9)^2 = 3.61$ and $(2.1)^2 = 4.41$, and thus the required interval 1.9 to 2.1, containing 2 not as an end-point, has been found. This example brings out the fact that statements about a function $f(x)$ in the neighbourhood of a number a are distinct from statements about the value of $f(x)$ when $x = a$. The production of an *interval*, throughout which the statement is true, is required. Thus the mere fact that $2^2 = 4$ does not by itself justify us in saying that in the *neighbourhood* of 2 the function x^2 is equal to 4. This statement would be untrue, because no interval can be produced with the required property. Also, the fact that $2^2 = 4$ does not by itself justify us in saying that in the *neighbourhood* of 2 the function x^2 approximates to 4 within the standard .5; although as a matter of fact, the statement has just been proved to be true.

If we understand the preceding ideas, we understand the foundations of modern mathematics. We shall recur to analogous ideas in the chapter on Series, and again in the chapter on the Differential Calculus. Meanwhile, we are now prepared to define "continuous functions." A function $f(x)$ is "continuous" at a value a of its argument, when in the neighbourhood of a its values approximate to $f(a)$ (*i.e.* to its value at a) within *every* standard of approximation.

This means that, whatever standard k be chosen, in the neighbourhood of a $f(x)$ approximates to $f(a)$ within the standard k. For example, x^2 is continuous at the value 2 of its argument, x, because however k be chosen we can always find an interval, which (i) contains 2 not as an end-point, and (ii) is such that the values of x^2 for arguments lying within it approximate to 4 (*i.e.* 2^2) within the standard k. Thus, suppose we choose the standard .1; now $(1.999)^2 = 3.996001$, and $(2.01)^2 = 4.0401$, and both these numbers differ from 4 by less than .1. Hence, within the interval 1.999 to 2.01 the values of x^2 approximate to 4 within the standard .1. Similarly an interval can be produced for any other standard which we like to try.

Take the example of the railway train. Its velocity is continuous as it passes the signal box, if whatever velocity you like to assign (say one-millionth of a mile per hour) an interval of time can be found extending before and after the instant of passing, such that at all instants within it the train's velocity differs from that with which the train passed the box by less than one-millionth of a mile per hour; and the same is true whatever other velocity be mentioned in the place of one-millionth of a mile per hour.

CHAPTER XII

PERIODICITY IN NATURE

THE whole life of Nature is dominated by the existence of periodic events, that is, by the existence of successive events so analogous to each other that, without any straining of language, they may be termed recurrences of the same event. The rotation of the earth produces the successive days. It is true that each day is different from the preceding days, however abstractly we define the meaning of a day, so as to exclude casual phenomena. But with a sufficiently abstract definition of a day, the distinction in properties between two days becomes faint and remote from practical interest; and each day may then be conceived as a recurrence of the phenomenon of one rotation of the earth. Again the path of the earth round the sun leads to the yearly recurrence of the seasons, and imposes another periodicity on all the operations of nature. Another less fundamental periodicity is provided by the phases of the moon. In modern civilized life, with its artificial light, these phases are of slight importance, but in ancient times, in climates where the days are burning and the skies clear, human life was apparently largely influenced by the existence of moonlight. Accordingly our divisions into weeks and months, with their religious associations, have spread over the European races from Syria and Mesopotamia, though independent observances following the moon's phases are found amongst most nations. It is, however, through the tides, and not through its phases of light and darkness, that the moon's periodicity has chiefly influenced the history of the earth.

Our bodily life is essentially periodic. It is dominated by the beatings of the heart, and the recurrence of breathing. The presupposition of periodicity is indeed fundamental to our very conception of life. We cannot imagine a course of nature in which, as events progressed, we should be unable to say: "This has happened before." The whole conception of experience as a guide to conduct would be absent. Men would always find themselves in new situations possessing no substratum of identity with anything in past history. The very means of measuring time as a quantity would be absent. Events might still be recognized as occurring in a series, so that some were earlier and others later. But we now go beyond this bare recognition. We can not only say that three events, A, B, C, occurred in this order, so that A came before B, and B before C; but also we can say that the length of time between the occurrences of A and B was twice as long as that between B and C. Now, quantity of time is essentially dependent on observing the number of natural recurrences which have intervened. We may say that the length of time between A and B was so many days, or so many months, or so many years, according to the type of recurrence to which we wish to appeal. Indeed, at the beginning of civilization, these three modes of measuring time were really distinct. It has been one of the first tasks of science among civilized or semi-civilized nations, to fuse them into one coherent measure. The full extent of this task must be grasped. It is necessary to determine, not merely what number of days (*e.g.* 365.25...) go to some one year, but also previously to determine that the same number of days do go to the successive years. We can imagine a world in which periodicities exist, but such that no two are coherent. In some years there might be 200 days and in others 350. The determination of the broad general consistency of the more important periodicities was the first step in natural science. This consistency arises from no abstract intuitive law of thought; it is merely an observed fact of nature guaranteed by experience. Indeed, so far is it from being a necessary law, that it is not even exactly true There are divergencies in every case. For some instances these divergencies are easily observed and are therefore immediately apparent. In other cases it requires the most refined observations and astronomical accuracy to make them apparent. Broadly speaking, all recurrences depending on living beings, such as the beatings of the heart, are subject in comparison with other recurrences to rapid variations. The great stable obvious recurrences—stable in the sense of mutually agreeing with great accuracy—are those depending on the motion of the earth as a whole, and on similar motions of the heavenly bodies.

We therefore assume that these astronomical recurrences mark out equal intervals of time. But how are we to deal with their discrepancies which the refined observations of astronomy detect? Apparently we are reduced to the arbitrary assumption that one or other of these sets of phenomena marks out equal times—*e.g.* that either all days are of equal length, or that all years are of equal length. This is not so: some assumptions must be made, but the assumption which underlies the whole procedure of the astronomers in determining the measure of time is that the laws of motion are exactly verified. Before explaining how this is done, it is interesting to observe that this relegation of the determination of the measure of time to the astronomers arises (as has been said) from the stable consistency of the recurrences with which they deal. If such a superior consistency had been noted among the recurrences characteristic of the human body, we should naturally have looked to the doctors of medicine for the regulation of our clocks.

In considering how the laws of motion come into the matter, note that two inconsistent modes of measuring time will yield different variations of velocity to the same body. For example, suppose we define an hour as one twenty-fourth of a day, and take the case of a train running uniformly for two hours at the rate of twenty miles per hour. Now take a grossly inconsistent measure of time, and suppose that it makes the first hour to be twice as long as the second hour. Then, according to this other measure of duration, the time of the train's run is divided into two parts, during each of which it has traversed the same distance, namely, twenty miles; but the duration of the first part is twice as long as that of the second part. Hence the velocity of the train has not been uniform, and on the average the velocity during the second period is twice that during the first period. Thus the question as to whether the train has been running uniformly or not entirely depends on the standard of time which we adopt.

Now, for all ordinary purposes of life on the earth, the various astronomical recurrences may be looked on as absolutely consistent; and, furthermore assuming their consistency, and thereby assuming the velocities and changes of velocities possessed by bodies, we find that the laws of motion, which have been considered above, are almost exactly verified. But only *almost* exactly when we come to some of the astronomical phenomena. We find, however, that by assuming slightly different velocities for the rotations and motions of the planets and stars, the laws would be exactly verified. This assumption is then made; and we have, in fact thereby, adopted a measure of time, which is indeed defined by reference to the astronomical phenomena, but not so as to be consistent with the uniformity of any one of them. But the broad fact remains that the uniform flow of time on which so much is based, is itself dependent on the observation of periodic events.

Even phenomena, which on the surface seem casual and exceptional, or, on the other hand, maintain themselves with a uniform persistency, may be due to the remote influence of periodicity. Take for example, the principle of resonance. Resonance arises when two sets of connected circumstances have the same periodicities. It is a dynamical law that the small vibrations of all bodies when left to themselves take place in definite times characteristic of the body. Thus a pendulum with a small swing always vibrates in some definite time, characteristic of its shape and distribution of weight and length. A more complicated body may have many ways of vibrating; but each of its modes of vibration will have its own peculiar "period." Those periods of vibration of a body are called its "free" periods. Thus a pendulum has but one period of vibration, while a suspension bridge will have many. We get a musical instrument, like a violin string, when the periods of vibration are all simple submultiples of the longest; *i.e.* if t seconds be the longest period, the others are $\frac{1}{2}t$, $\frac{1}{3}t$, and so on, where any of these smaller periods may be absent. Now, suppose we excite the vibrations of a body by a cause which is itself periodic; then, if the period of the cause is very nearly that of one of the periods of the body, that mode of vibration of the body is very violently excited; even although the magnitude of the exciting cause is small. This phenomenon is called "resonance." The general reason is easy to understand. Any one wanting to upset a rocking stone will push "in tune" with the oscillations of the stone, so as always to secure a favourable moment for a push. If the pushes are out of tune, some increase the oscillations, but others check them. But when they are in tune, after a time all the pushes are favourable. The word "resonance" comes from considerations of sound: but the phenomenon extends far beyond the region of sound. The laws of absorption and emission of light depend on it, the "tuning" of receivers for wireless telegraphy, the comparative importance of the influences of planets on each other's motion, the danger to a suspension bridge as troops march over it in step, and the excessive vibration of some ships under the rhythmical beat of their machinery at certain speeds. This coincidence of periodicities may produce steady phenomena when there is a constant association of the two periodic events, or it may produce violent and sudden outbursts when the association is fortuitous and temporary.

Again, the characteristic and constant periods of vibration mentioned above are the underlying causes of what appear to us as steady excitements of our senses. We work for hours in a steady light, or we listen to a steady unvarying sound. But, if modern science be correct, this steadiness has no counterpart in nature. The steady light is due to the impact on the eye of a countless number of periodic waves in a vibrating ether, and the steady sound to similar waves in a vibrating air. It is not our purpose here to explain the theory of light or the theory of sound. We have said enough to make it evident that one of the first steps necessary to make mathematics a fit instrument for the investigation of Nature is that it should be able to express the essential periodicity of things. If we have grasped this, we can understand the importance of the mathematical conceptions which we have next to consider, namely, periodic functions.

CHAPTER XIII

TRIGONOMETRY

Trigonometry did not take its rise from the general consideration of the periodicity of nature. In this respect its history is analogous to that of conic sections, which also had their origin in very particular ideas. Indeed, a comparison of the histories of the two sciences yields some very instructive analogies and contrasts. Trigonometry, like conic sections, had its origin among the Greeks. Its inventor was Hipparchus (born about 160 B.C.), a Greek astronomer, who made his observations at Rhodes. His services to astronomy were very great, and it left his hands a truly scientific subject with important results established, and the right method of progress indicated. Perhaps the invention of trigonometry was not the least of these services to the main science of his study. The next man who extended trigonometry was Ptolemy, the great Alexandrian astronomer, whom we have already mentioned. We now see at once the great contrast between conic sections and trigonometry. The origin of trigonometry was practical; it was invented because it was necessary for astronomical research. The origin of conic sections was purely theoretical. The only reason for its initial study was the abstract interest of the ideas involved. Characteristically enough conic sections were invented about 150 years earlier than trigonometry, during the very best period of Greek thought. But the importance of trigonometry, both to the theory and the application of mathematics, is only one of innumerable instances of the fruitful ideas which the general science has gained from its practical applications.

We will try and make clear to ourselves what trigonometry is, and why it should be generated by the scientific study of astronomy. In the first place: What are the measurements which can be made by an astronomer? They are measurements of time and measurements of angles. The astronomer may adjust a telescope (for it is easier to discuss the familiar instrument of modern astronomers) so that it can only turn about a fixed axis pointing east and west; the result is that the telescope can only point to the south, with a greater or less elevation of direction, or, if turned round beyond the zenith, point to the north. This is the transit instrument, the great instrument for the exact measurement of the times at which stars are due south or due north. But indirectly this instrument measures angles. For when the time elapsed between the transits of two stars has been noted, by the assumption of the uniform rotation of the earth, we obtain the angle through which the earth has turned in that period of time. Again, by other instruments, the angle between two stars can be directly measured. For if E is the eye of the astronomer, and EA and EB are the directions in which the stars are seen, it is easy to devise instruments which shall measure the angle AEB. Hence, when the astronomer is forming a survey of the heavens, he is, in fact, measuring angles so as to fix the relative directions of the stars and planets at any instant. Again, in the analogous problem of

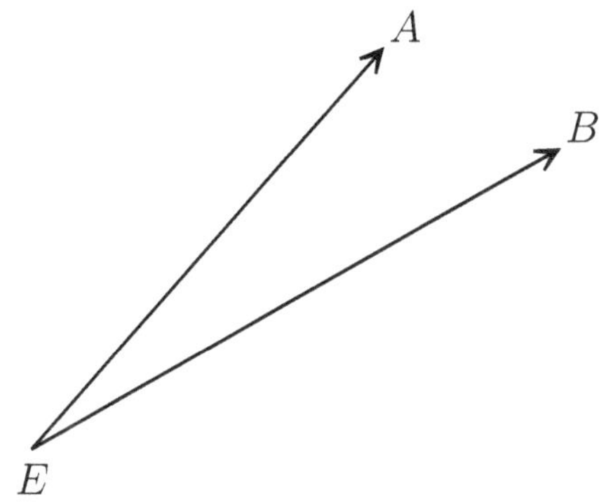

Fig. 22.

land-surveying, angles are the chief subject of measurements. The direct measurements of length are only rarely possible with any accuracy; rivers, houses, forests, mountains, and general irregularities of ground all get in the way. The survey of a whole country will depend only on one or two direct measurements of length, made with the greatest elaboration in selected places like Salisbury Plain. The main work of a survey is the measurement of angles. For example, A, B, and C will be conspicuous points in the district surveyed, say the tops of church towers. These points are visible each from the others. Then it is a very simple matter at A to measure the angle BAC, and at B to measure the angle ABC, and at C to measure the angle BCA. Theoretically, it is only necessary to measure two of these angles; for, by a well-known proposition in geometry, the sum of the three angles of a triangle amounts to two right-angles, so that when two

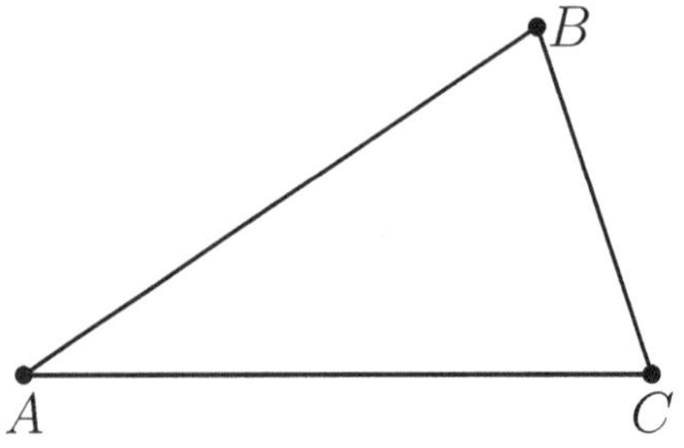

Fig. 23.

of the angles are known, the third can be deduced. It is better, however, in practice to measure all three, and then any small errors of observation can be checked. In the process of map-making a country is completely covered with triangles in this way. This process is called triangulation, and is the fundamental process in a survey.

Now, when all the angles of a triangle are known, the shape of the triangle is known—that is, the shape as distinguished from the size. We here come upon the great principle of geometrical similarity. The idea is very familiar to us in its practical applications. We are all familiar with the idea of a plan drawn to scale. Thus if the scale of a plan be an inch to a yard, a length of three inches in the plan means a length of three yards in the original. Also the shapes depicted in the plan are the shapes in the original, so that a right-angle in the original appears as a right-angle in the plan. Similarly in a map, which is only a plan of a country, the proportions of the lengths in the map are the proportions of the distances between the places indicated, and the directions in the map are the directions in the country. For example, if in the map one place is north-north-west of the other, so it is in reality; that is to say, in a map the angles are the same as in reality. Geometrical similarity may be defined thus: Two figures are similar (i) if to any point in one figure a point in the other figure corresponds, so that to every line there is a corresponding line, and to every angle a corresponding angle, and (ii) if the lengths of corresponding lines are in a fixed proportion, and the magnitudes of corresponding angles are the same. The fixed proportion of the lengths of corresponding lines in a map (or plan) and in the original is called the scale of the map. The scale should always be indicated on the margin of every map and plan. It has already been pointed out that two triangles whose angles are respectively equal are similar. Thus, if the two triangles ABC and DEF have the angles

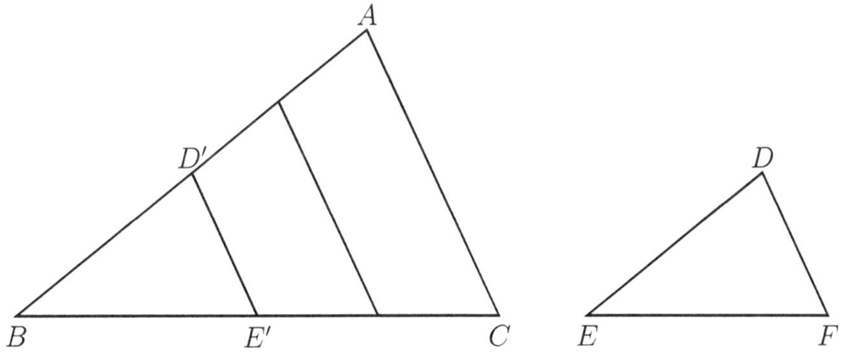

Fig. 24.

at A and D equal, and those at B and E, and those at C and F, then DE is to AB in the same proportion as EF is to BC, and as FD is to CA. But it is not true of other figures that similarity is guaranteed by the mere equality of angles. Take for example, the familiar cases of a rectangle and a square. Let $ABCD$ be a square, and $ABEF$ be a rectangle. Then all the corresponding angles are equal. But whereas the side AB of the square is equal to the side AB

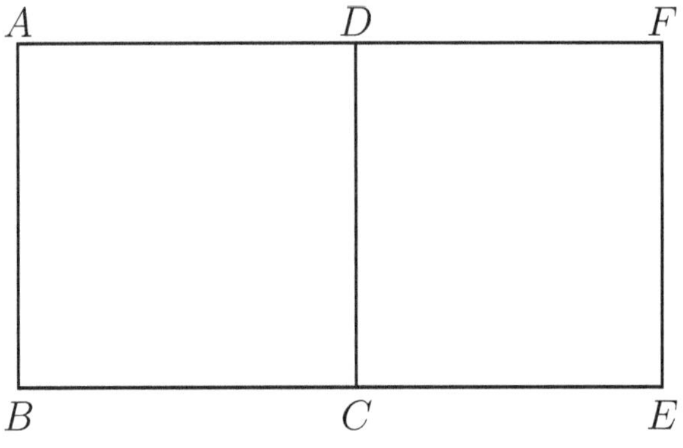

Fig. 25.

of the rectangle, the side BC of the square is about half the size of the side BE of the rectangle. Hence it is not true that the square $ABCD$ is similar to the rectangle $ABEF$. This peculiar property of the triangle, which is not shared by other rectilinear figures, makes it the fundamental figure in the theory of similarity. Hence in surveys, triangulation is the fundamental process; and hence also arises the word "trigonometry," derived from the two Greek words *trigonon* a triangle and *metria* measurement. The fundamental question from which trigonometry arose is this: Given the magnitudes of the angles of a triangle, what can be stated as to the relative magnitudes of the sides. Note that we say "*relative* magnitudes of the sides," since by the theory of similarity it is only the proportions of the sides which are known. In order to answer this question, certain functions of the magnitudes of an angle, considered as the argument, are introduced. In their origin these functions were got at by considering a right-angled triangle, and the magnitude of the angle was defined by the length of the arc of a circle. In modern elementary books, the fundamental position of the arc of the circle as defining the magnitude of the angle has been pushed somewhat to the background, not to the advantage either of theory or clearness of explanation. It must first be noticed that, in relation to similarity, the circle holds the same fundamental position among curvilinear figures, as does the triangle among rectilinear figures. Any two circles are similar figures; they only differ in scale. The lengths of the circumferences of two circles, such as APA' and $A_1P_1A'_1$ in the fig. 26 are in proportion to the lengths of their radii. Furthermore, if the two circles have the same centre O, as do the two circles in fig. 26, then the arcs AP and A_1P_1 intercepted by the arms of any angle AOP, are also in proportion to their radii. Hence the ratio of the length of the arc AP to the length of the radius OP, that is $\frac{\text{arc } AP}{\text{radius } OP}$ is a number which is quite independent of the length OP,

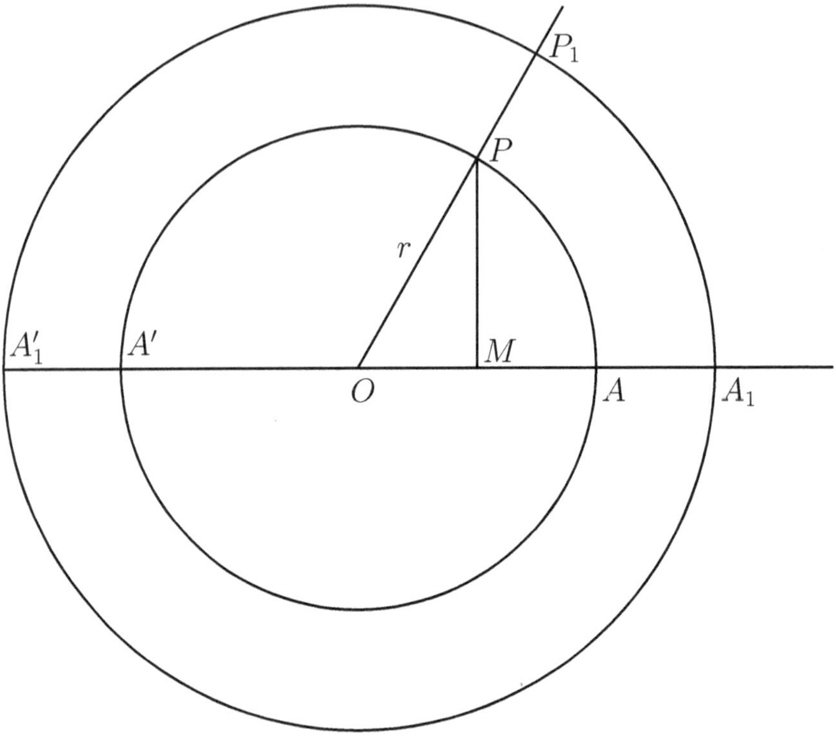

Fig. 26.

and is the same as the fraction $\frac{\text{arc } A_1P_1}{\text{radius } OP_1}$. This fraction of "arc divided by radius" is the proper theoretical way to measure the magnitude of an angle; for it is dependent on no arbitrary unit of length, and on no arbitrary way of dividing up any arbitrarily assumed angle, such as a right-angle. Thus the fraction $\frac{AP}{OA}$ represents the magnitude of the angle AOP. Now draw PM perpendicularly to OA. Then the Greek mathematicians called the line PM the sine of the arc AP, and the line OM the cosine of the arc AP. They were well aware that the importance of the relations of these various lines to each other was dependent on the theory of similarity which we have just expounded. But they did not make their definitions express the properties which arise from this theory. Also they had not in their heads the modern general ideas respecting functions as correlating pairs of variable numbers, nor in fact were they aware of any modern conception of algebra and algebraic analysis. Accordingly, it was natural to them to think merely of the relations between certain lines in a diagram. For us the case is different: we wish to embody our more powerful ideas.

Hence, in modern mathematics, instead of considering the arc AP, we consider the fraction $\frac{AP}{OP}$, which is a number the same for all lengths of OP; and, instead of considering the lines PM and OM, we consider the fractions $\frac{PM}{OP}$ and $\frac{OM}{OP}$, which again are numbers not dependent on the length of OP, *i.e.* not dependent on the scale of our diagrams. Then we define the number $\frac{PM}{OP}$ to be the *sine* of the number $\frac{PA}{OP}$, and the number $\frac{OM}{OP}$ to be the *cosine* of the number $\frac{PA}{OP}$. These fractional forms are clumsy to print; so let us put u for the fraction $\frac{AP}{OP}$, which represents the magnitude of the angle AOP, and put v for the fraction $\frac{PM}{OM}$, and w for the fraction $\frac{OM}{OP}$. Then u, v, w, are numbers, and, since we are talking of *any* angle AOP, they are variable numbers. But a correlation exists between their magnitudes, so that when u (*i.e.* the angle AOP) is given the magnitudes of v and w are definitely determined. Hence v and w are functions of the argument u. We have called v the *sine* of u, and w the *cosine* of u. We wish to adapt the general functional notation $y = f(x)$ to these special cases: so in modern mathematics we write "sin" for "f" when we want to indicate the special function of "sine," and "cos" for "f" when we want to indicate the special function of "cosine." Thus, with the above meanings for u, v, w, we get

$$v = \sin u, \quad \text{and} \quad w = \cos u,$$

where the brackets surrounding the x in $f(x)$ are omitted for the special functions. The meaning of these functions sin and cos as correlating the pairs of numbers u and v, and u and w is, that the functional relations are to be found by constructing (*cf.* fig. 26) an angle AOP, whose measure "AP divided by OP" is equal to u, and that then v is the number given by "PM divided by OP" and w is the number given by "OM divided by OP."

It is evident that without some further definitions we shall get into difficulties when the number u is taken too large. For then the arc AP may be greater than one-quarter of the circumference of the circle, and the point M (*cf.* figs. 26 and 27) may fall between O and A' and not between O and A. Also P may be below the line AOA' and not above it as in fig. 26. In order to get over this difficulty we have recourse to the ideas and conventions of coordinate geometry in making our complete definitions of the sine and cosine. Let one arm OA of the angle be the axis OX, and produce the axis backwards to obtain its negative part OX'. Draw the other axis YOY' perpendicular to it. Let any point P at a distance r from O have coordinates x and y. These coordinates are both positive in the first "quadrant" of the plan, *e.g.* the coordinates x and y of P in fig. 27. In the other quadrants, either one or both of the coordinates are negative, for example, x' and y for P', and x' and y' for P'', and x and y' for P''' in fig. 27, where x' and y' are both negative numbers. The positive angle POA is the arc AP divided by r, its sine is $\frac{y}{r}$ and its cosine is $\frac{x}{r}$; the positive angle AOP' is the arc ABP' divided by r, its sine is $\frac{y}{r}$ and cosine $\frac{x'}{r}$; the positive angle AOP'' is the arc $ABA'P''$ divided by r, its sine is $\frac{y'}{r}$ and its cosine is $\frac{x'}{r}$; the positive angle AOP''' is the arc $ABA'B'P'''$ divided by r, its sine is $\frac{y'}{r}$ and its cosine is $\frac{x}{r}$.

But even now we have not gone far enough. For suppose we choose u to be a number greater than the ratio of the whole circumference of the circle to its radius. Owing to the similarity of all circles this ratio is the same for all cir-

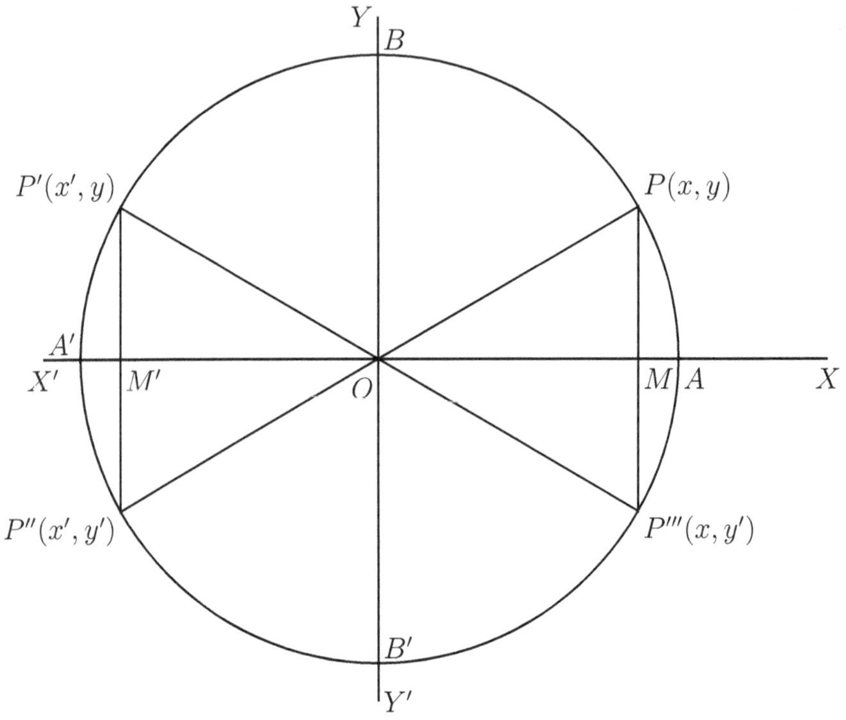

Fig. 27.

cles. It is always denoted in mathematics by the symbol 2π, where π is the Greek form of the letter p and its name in the Greek alphabet is "pi." It can be proved that π is an incommensurable number, and that therefore its value cannot be expressed by any fraction, or by any terminating or recurring decimal. Its value to a few decimal places is 3.14159; for many purposes a sufficiently accurate approximate value is $\frac{22}{7}$. Mathematicians can easily calculate π to any degree of accuracy required, just as $\sqrt{2}$ can be so calculated. Its value has been actually given to 707 places of decimals. Such elaboration of calculation is merely a curiosity, and of no practical or theoretical interest. The accurate determination of π is one of the two parts of the famous problem of squaring the circle. The other part of the problem is, by the theoretical methods of pure geometry to describe a straight line equal in length to the circumference. Both parts of the problem are now known to be impossible; and the insoluble problem has now lost all special practical or theoretical interest, having become absorbed in wider ideas.

After this digression on the value of π, we now return to the question of the general definition of the magnitude of an angle, so as to be able to produce an angle corresponding to any value u. Suppose a moving point, Q, to start from A on OX (*cf.* fig. 27), and to rotate in the positive direction (anti-clockwise, in the figure considered) round the circumference of the circle for any number of times, finally resting at any point, *e.g.* at P or P' or P'' or P'''. Then the total length of the curvilinear circular path traversed, divided by the radius of the circle, r, is the generalized definition of a positive angle of *any* size. Let x, y be the coordinates of the point in which the point Q rests, *i.e.* in one of the four alternative positions mentioned in fig. 27; x and y (as here used) will either be x and y, or x' and y, or x' and y', or x and y'. Then the sign of this generalized angle is $\frac{y}{r}$ and its cosine is $\frac{x}{r}$. With these definitions the functional relations

$v = \sin u$ and $w = \cos u$, are at last defined for all positive real values of u. For negative values of u we simply take rotation of Q in the opposite (clockwise) direction; but it is not worth our while to elaborate further on this point, now that the general method of procedure has been explained.

These functions of sine and cosine, as thus defined, enable us to deal with the problems concerning the triangle from which Trigonometry took its rise. But we are now in a position to relate Trigonometry to the wider idea of Periodicity of which the importance was explained in the last chapter. It is easy to see that the functions $\sin u$ and $\cos u$ are periodic functions of u. For consider the position, P (in fig. 27), of a moving point, Q, which has started from A and revolved round the circle. This position, P, marks the angles $\frac{\text{arc } AP}{r}$, and $2\pi + \frac{\text{arc } AP}{r}$, and $4\pi + \frac{\text{arc } AP}{r}$, and $6\pi + \frac{\text{arc } AP}{r}$, and so on indefinitely. Now, all these angles have the same sine and cosine, namely, $\frac{y}{r}$ and $\frac{x}{r}$. Hence it is easy to see that, if u be chosen to have any value, the arguments u and $2\pi + u$, and $4\pi + u$, and $6\pi + u$, and $8\pi + u$ and so on indefinitely, have all the same values for the corresponding sines and cosines. In other words,

$$\sin u = \sin(2\pi + u) = \sin(4\pi + u) = \sin(6\pi + u) = \text{etc.};$$
$$\cos u = \cos(2\pi + u) = \cos(4\pi + u) = \cos(6\pi + u) = \text{etc.}$$

This fact is expressed by saying that $\sin u$ and $\cos u$ are periodic functions with their period equal to 2π.

The graph of the function $y = \sin x$ (notice that we now abandon v and u for the more familiar y and x) is shown in fig. 28. We take on the axis of x any arbitrary length at pleasure to represent the number π, and on the axis of y any arbitrary length at pleasure to represent the number 1. The numerical values of the sine and cosine can never exceed unity. The recurrence of the figure after periods of 2π will be noticed. This graph represents the simplest style of periodic function, out of which all others are constructed. The cosine gives nothing fundamentally different from the sine. For it is easy to prove that $\cos x = \sin(x + \frac{\pi}{2})$; hence it can be seen that the graph of $\cos x$ is simply fig. 28 modified by drawing the axis of OY through the point on OX marked $\frac{\pi}{2}$, instead of drawing it in its actual position on the figure.

It is easy to construct a 'sine' function in which the period

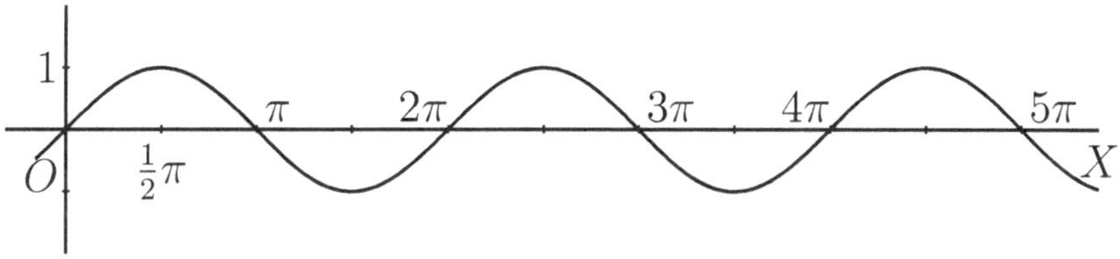

Fig. 28.

has any assigned value a. For we have only to write

$$y = \sin \frac{2\pi x}{a},$$

and then

$$\sin \frac{2\pi(x+a)}{a} = \sin\left(\frac{2\pi x}{a} + 2\pi\right) = \sin \frac{2\pi x}{a}.$$

Thus the period of this new function is now a. Let us now give a general definition of what we mean by a periodic function. The function $f(x)$ is periodic, with the period a, if (i) for *any* value of x we have $f(x) = f(x+a)$, and (ii) there is no number b smaller than a such that for *any* value of x, $f(x) = f(x+b)$.

The second clause is put into the definition because when we have $\sin \frac{2\pi x}{a}$, it is not only periodic in the period a, but also in the periods $2a$ and $3a$, and so on; this arises since

$$\sin \frac{2\pi(x+3a)}{a} = \sin\left(\frac{2\pi x}{a} + 6\pi\right) = \sin \frac{2\pi x}{a}.$$

So it is the smallest period which we want to get hold of and call *the* period of the function. The greater part of the abstract theory of periodic functions and the whole of the applications of the theory to Physical Science are dominated by an important theorem called Fourier's Theorem; namely that, if $f(x)$ be a periodic function with the period a and if $f(x)$ also satisfies certain conditions, which practically are always presupposed in functions suggested by natural phenomena, then $f(x)$ can be written as the sum of a set of terms in the form

$$c_0 + c_1 \sin\left(\frac{2\pi x}{a} + e_1\right) + c_2 \sin\left(\frac{4\pi x}{a} + e_2\right) + c_3 \sin\left(\frac{6\pi x}{a} + e_3\right) + \text{etc.}$$

In this formula c_0, c_1, c_2, c_3, etc., and also e_1, e_2, e_3, etc., are constants, chosen so as to suit the particular function. Again we have to ask, How many terms have to be chosen? And here a new difficulty arises: for we can prove that, though in some particular cases a definite number will do, yet in general all we can do is to approximate as closely as we like to the value of the function by taking more and more terms. This process of gradual approximation brings us to the consideration of the theory of infinite series, an essential part of mathematical theory which we will consider in the next chapter.

The above method of expressing a periodic function as a sum of sines is called the "harmonic analysis" of the function. For example, at any point on the sea coast the tides rise and fall periodically. Thus at a point near the Straits of Dover there will be two daily tides due to the rotation of the earth. The daily rise and fall of the tides are complicated by the fact that there are two tidal waves, one coming up the English Channel, and the other which has swept round the North of Scotland, and has then come southward down the North Sea. Again some high tides are higher than others: this is due to the fact that the Sun has also a tide-generating influence as well as the Moon. In this way monthly and other periods are introduced. We leave out of account the exceptional influence of winds which cannot be foreseen. The general problem of the harmonic analysis of the tides is to find sets of terms like those in the expression on page 157 above, such that each set will give with approximate accuracy the contribution of the tide-generating influences of one "period" to the height of the tide at any instant. The argument x will therefore be the *time* reckoned from any convenient commencement.

Again, the motion of vibration of a violin string is submitted to a similar harmonic analysis, and so are the vibrations of the ether and the air, corresponding respectively to waves of light and waves of sound. We are here in the presence of one of the fundamental processes of mathematical physics—namely, nothing less than its general method of dealing with the great natural fact of Periodicity.

CHAPTER XIV

SERIES

NO part of Mathematics suffers more from the triviality of its initial presentation to beginners than the great subject of series. Two minor examples of series, namely arithmetic and geometric series, are considered; these examples are important because they are the simplest examples of an important general theory. But the general ideas are never disclosed; and thus the examples, which exemplify nothing, are reduced to silly trivialities.

The general mathematical idea of a series is that of a set of things ranged in order, that is, in sequence; This meaning is accurately represented in the common use of the term. Consider for example, the series of English Prime Ministers during the nineteenth century, arranged in the order of their first tenure of that office within the century. The series commences with William Pitt, and ends with Lord Rosebery, who, appropriately enough, is the biographer of the first member. We might have considered other serial orders for the arrangement of these men; for example, according to their height or their weight. These other suggested orders strike us as trivial in connection with Prime Ministers, and would not naturally occur to the mind; but abstractly they are just as good orders as any other. When one order among terms is very much more important or more obvious than other orders, it is often spoken of as *the* order of those terms. Thus *the* order of the integers would always be taken to mean their order as arranged in order of magnitude. But of course there is an indefinite number of other ways of arranging them. When the number of things considered is finite, the number of ways of arranging them in order is called the number of their permutations. The number of permutations of a set of n things, where n is some finite integer, is

$$n \times (n-1) \times (n-2) \times (n-3) \times \ldots \times 4 \times 3 \times 2 \times 1,$$

that is to say, it is the product of the first n integers; this product is so important in mathematics that a special symbolism, is used for it, and it is always written '$n!$.' Thus, $2! = 2 \times 1 = 2$, and $3! = 3 \times 2 \times 1 = 6$, and $4! = 4 \times 3 \times 2 \times 1 = 24$, and $5! = 5 \times 4 \times 3 \times 2 \times 1 = 120$. As n increases, the value of $n!$ increases very quickly; thus $100!$ is a hundred times as large as $99!$.

It is easy to verify in the case of small values of n that $n!$ is the number of ways of arranging n things in order. Thus consider two things a and b; these are capable of the two orders ab and ba, and $2! = 2$.

Again, take three things a, b, and c; these are capable of the six orders, abc, acb, bac, bca, cab, cba, and $3! = 6$. Similarly for the twenty-four orders in which four things a, b, c, and d, can be arranged.

When we come to the infinite sets of things—like the sets of all the integers, or all the fractions, or all the real numbers for instance—we come at once upon the complications of the theory of order-types. This subject was touched upon in Chapter VI. in considering the possible orders of the integers, and of the fractions, and of the real numbers. The whole question of order-types forms a comparatively new branch of mathematics of great importance. We shall not consider it any further. All the infinite series which we consider now are of the same order-type as the integers arranged in ascending order of magnitude, namely, with a first term, and such that each term has a couple of next-door neighbours, one on either side, with the exception of the first term which has, of course, only one next-door neighbour. Thus, if m be any integer (not zero), there will be always an mth term. A series with a finite number of terms (say n terms) has the same characteristics as far as next-door neighbours are concerned as an infinite series; it only differs from infinite series in having a last term, namely, the nth.

The important thing to do with a series of numbers—using for the future "series" in the restricted sense which has just been mentioned—is to add its successive terms together.

Thus if u_1, u_2, u_3, $\ldots$, u_n, $\ldots$ are respectively the 1st, 2nd, 3rd, 4th, $\ldots$, nth, $\ldots$ terms of a series of numbers, we form successively the series u_1, $u_1 + u_2$, $u_1 + u_2 + u_3$, $u_1 + u_2 + u_3 + u_4$, and so on; thus the sum of the 1st n terms may be written

$$u_1 + u_2 + u_3 + \cdots + u_n.$$

If the series has only a finite number of terms, we come at last in this way to the sum of the whole series of terms. But, if the series has an infinite number of terms, this process of successively forming the sums of the terms never terminates; and in this sense there is no such thing as the sum of an infinite series.

But why is it important successively to add the terms of a series in this way? The answer is that we are here symbolizing the fundamental mental process of approximation. This is a process which has significance far beyond the regions of mathematics. Our limited intellects cannot deal with complicated material all at once, and our method of arrangement is that of approximation. The statesman in framing his speech puts the dominating issues first and lets the details fall naturally into their subordinate places. There is, of course, the converse artistic method of preparing the imagination by the presentation of subordinate or special details, and then gradually rising to a crisis. In either way the process is one of gradual summation of effects; and this is exactly what is done by the successive summation of the terms of a series. Our ordinary method of stating numbers is such a process of gradual summation, at least, in the case of large numbers. Thus $568,213$ presents itself to the mind as:—

$$500,000 + 60,000 + 8,000 + 200 + 10 + 3.$$

In the case of decimal fractions this is so more avowedly. Thus 3.14159 is:—

$$3 + \frac{1}{10} + \frac{4}{100} + \frac{1}{1000} + \frac{5}{10000} + \frac{9}{100000}.$$

Also, 3 and $3+\frac{1}{10}$, and $3+\frac{1}{10}+\frac{4}{100}$, and $3+\frac{1}{10}+\frac{4}{100}+\frac{1}{1000}$, and $3+\frac{1}{10}+\frac{4}{100}+\frac{1}{1000}+\frac{5}{10000}$ are successive approximations to the complete result 3.14159. If we read 568, 213 backwards from right to left, starting with the 3 units, we read it in the artistic way, gradually preparing the mind for the crisis of 500, 000.

The ordinary process of numerical multiplication proceeds by means of the summation of a series, Consider the computation

$$\begin{array}{r} 342 \\ \underline{658} \\ 2736 \\ 1710 \\ \underline{2052} \\ 225036 \end{array}$$

Hence the three lines to be added form a series of which the first term is the upper line. This series follows the artistic method of presenting the most important term last, not from any feeling for art, but because of the convenience gained by keeping a firm hold on the units' place, thus enabling us to omit some 0's, formally necessary.

But when we approximate by gradually adding the successive terms of an infinite series, what are we approximating to? The difficulty is that the series has no "sum" in the straightforward sense of the word, because the operation of adding together its terms can never be completed. The answer is that we are approximating to the *limit* of the summation of the series, and we must now proceed to explain what the "limit" of a series is.

The summation of a series approximates to a limit when the sum of any number of its terms, provided the number be large enough, is as nearly equal to the limit as you care to approach. But this description of the meaning of approximating to a limit evidently will not stand the vigorous scrutiny of modern mathematics. What is meant by *large enough*, and by *nearly equal*, and by *care to approach*? All these vague phrases must be explained in terms of the simple abstract ideas which alone are admitted into pure mathematics.

Let the successive terms of the series be $u_1, u_2, u_3, u_4, \ldots, u_n$, etc., so that u_n is the nth term of the series. Also let s_n be the sum of the 1st n terms, whatever n may be. So that:—

$$s_1 = u_1, \quad s_2 = u_1 + u_2, \quad s_3 = u_1 + u_2 + u_3, \quad \text{and}$$
$$s_n = u_1 + u_2 + u_3 + \cdots + u_n.$$

Then the terms $s_1, s_2, s_3, \ldots, s_n, \ldots$ form a new series, and the formation of this series is the process of summation of the original series. Then the "approximation" of the *summation* of the original series to a "limit" means the "approximation of the *terms* of this new series to a limit." And we have now to explain what we mean by the approximation to a limit of the terms of a series.

Now, remembering the definition (given in Chapter XII.) of a *standard of approximation*, the idea of a limit means this: l is the limit of the terms of the series $s_1, s_2, s_3, \ldots, s_n, \ldots$, if, corresponding to each real number k, taken as a standard of approximation, a term s_n of the series can be found so that all succeeding terms (*i.e.* s_{n+1}, s_{n+2}, etc.) approximate to l within that standard of approximation. If another smaller standard k^1 be chosen, the term s_n may be too early in the series, and a later term s_m with the above property will then be found.

If this property holds, it is evident that as you go along to series $s_1, s_2, s_3, \ldots, s_n, \ldots$ from left to right, after a time you come to terms *all* of which are nearer to l than any number which you may like to assign. In other words you approximate to l as closely as you like. The close connection of this definition of the limit of a series with the definition of a continuous function given in Chapter XI. will be immediately perceived.

Then coming back to the original series $u_1, u_2, u_3, \ldots, u_n, \ldots$, the limit of the terms of the series $s_1, s_2, s_3, \ldots, s_n, \ldots$, is called the "sum to infinity" of the original series. But it is evident that this use of the word "sum" is very artificial, and we must not assume the analogous properties to those of the ordinary sum of a finite number of terms without some special investigation.

Let us look at an example of a "sum to infinity." Consider the recurring decimal .1111 .... This decimal is merely a way of symbolizing the "sum to infinity" of the series .1, .01, .001, .0001, etc. The corresponding series found by summation is $s_1 = .1$, $s_2 = .11$, $s_3 = .111$, $s_4 = .1111$, etc. The limit of the terms of this series is $\frac{1}{9}$; this is easy to see by simple division, for

$$\tfrac{1}{9} = .1 + \tfrac{1}{90} = .11 + \tfrac{1}{900} = .111 + \tfrac{1}{9000} = \text{etc.}$$

Hence, if $\frac{3}{17}$ is given (the k of the definition), .1 and *all* succeeding terms differ from $\frac{1}{9}$ by less than $\frac{3}{17}$; if $\frac{1}{1000}$ is given (another choice for the k of the definition), .111 and all succeeding terms differ from $\frac{1}{9}$ by less than $\frac{1}{1000}$; and so on, whatever choice for k be made.

It is evident that nothing that has been said gives the slightest idea as to how the "sum to infinity" of a series is to be found. We have merely stated the conditions which such a number is to satisfy. Indeed, a general method for finding in all cases the sum to infinity of a series is intrinsically out of the question, for the simple reason that such a "sum," as here defined, does not always exist. Series which possess a sum to infinity are called *convergent*, and those which do not possess a sum to infinity are called *divergent*.

An obvious example of a divergent series is 1, 2, 3, ..., n ..., *i.e.* the series of integers in their order of magnitude. For whatever number l you try to take as its sum to infinity, and whatever standard of approximation k you choose, by taking enough terms of the series you can always make their sum differ from l by more than k. Again, another example of a divergent series is 1, 1, 1, etc., *i.e.* the series of which each term is equal to 1. Then the sum of n terms is n, and this sum grows without limit as n increases. Again, another example of a divergent series is 1, −1, 1, −1, 1, −1, etc., *i.e.* the series in which the terms are alternately 1 and -1. The sum of an odd number of terms is 1, and of an even number of terms is 0. Hence the terms of the series s_1, s_2, s_3, ..., s_n, ... do not approximate to a limit, although they do not increase without limit.

It is tempting to suppose that the condition for $u_1, u_2, \ldots, u_n, \ldots$ to have a sum to infinity is that u_n should decrease indefinitely as n increases. Mathematics would be a much easier science than it is, if this were the case. Unfortunately the supposition is not true.

For example the series

$$1, \quad \frac{1}{2}, \quad \frac{1}{3}, \quad \frac{1}{4}, \quad \ldots, \quad \frac{1}{n}, \quad \ldots$$

is divergent. It is easy to see that this is the case; for consider the sum of n terms beginning at the $(n+1)$th term. These n terms are $\frac{1}{n+1}, \frac{1}{n+2}, \frac{1}{n+3}, \ldots, \frac{1}{2n}$: there are n of them and $\frac{1}{2n}$ is the least among them. Hence their sum is greater than n times $\frac{1}{2n}$, *i.e.* is greater than $\frac{1}{2}$. Now, without altering the sum to infinity, if it exist, we can add together neighbouring terms, and obtain the series

$$1, \quad \tfrac{1}{2}, \quad \tfrac{1}{3}+\tfrac{1}{4}, \quad \tfrac{1}{5}+\tfrac{1}{6}+\tfrac{1}{7}+\tfrac{1}{8}, \quad \text{etc.,}$$

that is, by what has been said above, a series whose terms after the 2nd are greater than those of the series,

$$1, \quad \tfrac{1}{2}, \quad \tfrac{1}{2}, \quad \tfrac{1}{2}, \quad \text{etc.,}$$

where all the terms after the first are equal. But this series is divergent. Hence the original series is divergent.*

This question of divergency shows how careful we must be in arguing from the properties of the sum of a finite number of terms to that of the sum of an infinite series. For the most elementary property of a finite number of terms is that of course they possess a sum: but even this fundamental property is not necessarily possessed by an infinite series. This caution merely states that we must not be misled by the suggestion of the technical term "*sum* of an infinite series." It is usual to indicate the sum of the infinite series

$$u_1, \quad u_2, \quad u_3, \ \ldots, \quad u_n, \ \ldots$$

by

$$u_1 + u_2 + u_3 + \cdots + u_n + \ldots.$$

We now pass on to a generalization of the idea of a series, which mathematics, true to its method, makes by use of the variable. Hitherto, we have only contemplated series in which each definite term was a definite number. But equally well we can generalize, and make each term to be some mathematical expression containing a variable x. Thus we may consider the series $1, x, x^2, x^3, \ldots, x^n, \ldots$, and the series

$$x, \quad \frac{x^2}{2}, \quad \frac{x^3}{3}, \ \ldots, \quad \frac{x^n}{n}, \ \ldots.$$

* *Cf.* Note C, p. 208.

In order to symbolize the general idea of any such function, conceive of a function of x, $f_n(x)$ say, which involves in its formation a variable integer n, then, by giving n the values 1, 2, 3, etc., in succession, we get the series

$$f_1(x), \quad f_2(x), \quad f_3(x), \ \ldots, \quad f_n(x), \ldots.$$

Such a series may be convergent for some values of x and divergent for others. It is, in fact, rather rare to find a series involving a variable x which is convergent for all values of x,—at least in any particular instance it is very unsafe to assume that this is the case. For example, let us examine the simplest of all instances, namely, the "geometrical" series

$$1, \quad x, \quad x^2, \quad x^3, \ \ldots, \quad x^n, \ \ldots.$$

The sum of n terms is given by

$$s_n = 1 + x + x^2 + x^3 + \cdots + x^n.$$

Now multiply both sides by x and we get

$$xs_n = x + x^2 + x^3 + x^4 + \cdots + x^n + x^{n+1}.$$

Now subtract the last line from the upper line and we get

$$s_n(1 - x) = s_n - xs_n = 1 - x^{n+1},$$

and hence (if x be not equal to 1)

$$s_n = \frac{1 - x^{n+1}}{1 - x} = \frac{1}{1 - x} - \frac{x^{n+1}}{1 - x}.$$

Now if x be numerically less than 1, for sufficiently large values of n, $\frac{x^{n+1}}{1-x}$ is always numerically less than k, however k be chosen. Thus, if x be numerically less than 1, the series $1,\ x,\ x^2,\ \ldots,\ x^n,\ \ldots$ is convergent, and $\frac{1}{1-x}$ is its limit. This statement is symbolized by

$$\frac{1}{1-x} = 1 + x + x^2 + \cdots + x^n + \ldots, \qquad (-1 < x < 1).$$

But if x is numerically greater than 1, or numerically equal to 1, the series is divergent. In other words, if x lie between -1 and $+1$, the series is convergent; but if x be equal to -1 or $+1$, or if x lie outside the interval -1 to $+1$, then the series is divergent. Thus the series is convergent at all "points" within the interval -1 to $+1$, exclusive of the end points.

At this stage of our enquiry another question arises. Suppose that the series

$$f_1(x) + f_2(x) + f_3(x) + \cdots + f_n(x) + \ldots$$

is convergent for all values of x lying within the interval a to b, *i.e.* the series is convergent for any value of x which is greater than a and less than b. Also, suppose we want to be sure that in approximating to the limit we add together enough terms to come within some standard of approximation k. Can we always state some number of terms, say n, such that, if we take n or more terms to form the sum, then *whatever* value x has within the interval we have satisfied the desired standard of approximation?

Sometimes we can and sometimes we cannot do this for each value of k. When we can, the series is called uniformly convergent throughout the interval, and when we cannot do so, the series is called non-uniformly convergent throughout the interval. It makes a great difference to the properties of a series whether it is or is not uniformly convergent through an interval. Let us illustrate the matter by the simplest example and the simplest numbers.

Consider the geometric series

$$1 + x + x^2 + x^3 + \cdots + x^n + \ldots.$$

It is convergent throughout the interval -1 to $+1$, excluding the end values $x = \pm 1$.

But it is not uniformly convergent throughout this interval. For if $s_n(x)$ be the sum of n terms, we have proved that the difference between $s_n(x)$ and the limit $\frac{1}{1-x}$ is $\frac{x^{n+1}}{1-x}$. Now suppose n be any given number of terms, say 20, and let k be any assigned standard of approximation, say .001. Then, by taking x near enough to $+1$ or near enough to -1, we can make the numerical value of $\frac{x^{21}}{1-x}$ to be greater than .001. Thus 20 terms will not do over the whole interval, though it is more than enough over some parts of it.

The same reasoning can be applied whatever other number we take instead of 20, and whatever standard of approximation instead of .001. Hence the geometric series

$1 + x + x^2 + x^3 + \cdots + x^n + \ldots$ is non-uniformly convergent over its *whole* interval of convergence -1 to $+1$. But if we take any smaller interval lying at both ends within the interval -1 to $+1$, the geometric series is uniformly convergent within it. For example, take the interval 0 to $+\frac{1}{10}$. Then any value for n which makes $\dfrac{x^{n+1}}{1-x}$ numerically less than k *at* these limits for x also serves for all values of x between these limits, since it so happens that $\dfrac{x^{n+1}}{1-x}$ diminishes in numerical value as x diminishes in numerical value. For example, take $k = .001$; then, putting $x = \frac{1}{10}$, we find:—

$$\text{for } n = 1, \quad \frac{x^{n+1}}{1-x} = \frac{(\frac{1}{10})^2}{1-\frac{1}{10}} = \tfrac{1}{90} = .0111\ldots,$$

$$\text{for } n = 2, \quad \frac{x^{n+1}}{1-x} = \frac{(\frac{1}{10})^3}{1-\frac{1}{10}} = \tfrac{1}{900} = .00111\ldots,$$

$$\text{for } n = 3, \quad \frac{x^{n+1}}{1-x} = \frac{(\frac{1}{10})^4}{1-\frac{1}{10}} = \tfrac{1}{9000} = .000111\ldots.$$

Thus three terms will do for the whole interval, though, of course, for some parts of the interval it is more than is necessary. Notice that, because $1+x+x^2+\cdots+x^n+\ldots$ is convergent (though not uniformly) throughout the interval -1 to $+1$, for each value of x in the interval some number of terms n can be found which will satisfy a desired standard of approximation; but, as we take x nearer and nearer to either end value $+1$ or -1, larger and larger values of n have to be employed.

It is curious that this important distinction between uniform and non-uniform convergence was not published till 1847 by Stokes—afterwards, Sir George Stokes—and later, independently in 1850 by Seidel, a German mathematician.

The critical points, where non-uniform convergence comes in, are not necessarily at the limits of the interval throughout which convergence holds. This is a speciality belonging to the geometric series.

In the case of the geometric series $1+x+x^2+\cdots+x^n+\ldots$, a simple algebraic expression $\frac{1}{1-x}$ can be given for its limit in its interval of convergence. But this is not always the case. Often we can prove a series to be convergent within a certain interval, though we know nothing more about its limit except that it is the limit of the series. But this is a very good way of defining a function; *viz.*. as the limit of an infinite convergent series, and is, in fact, the way in which most functions are, or ought to be, defined.

Thus, the most important series in elementary analysis is

$$1+x+\frac{x^2}{2!}+\frac{x^3}{3!}+\cdots+\frac{x^n}{n!}+\ldots,$$

where $n!$ has the meaning defined earlier in this chapter. This series can be proved to be absolutely convergent for *all* values of x, and to be uniformly convergent within any interval which we like to take. Hence it has all the comfortable mathematical properties which a series should have. It is called the exponential series. Denote its sum to infinity by $\exp x$. Thus, by definition,

$$\exp x = 1 + x + \frac{x^2}{2!} + \frac{x^3}{3!} + \cdots + \frac{x^n}{n!} + \ldots.$$

$\exp x$ is called the exponential function.

It is fairly easy to prove, with a little knowledge of elementary mathematics, that

$$(A) \qquad (\exp x) \times (\exp y) = \exp(x + y).$$

In other words that

$$(\exp x) \times (\exp y)$$
$$= 1 + (x + y) + \frac{(x+y)^2}{2!} + \frac{(x+y)^3}{3!} + \cdots + \frac{(x+y)^n}{n!} + \ldots.$$

This property (A) is an example of what is called an addition-theorem. When any function [say $f(x)$] has been defined, the first thing we do is to try to express $f(x + y)$ in terms of known functions of x only, and known functions of y only. If we can do so, the result is called an addition-theorem. Addition-theorems play a great part in mathematical analysis. Thus the addition-theorem for the sine is given by

$$\sin(x + y) = \sin x \cos y + \cos x \sin y,$$

and for the cosine by

$$\cos(x + y) = \cos x \cos y - \sin x \sin y.$$

As a matter of fact the best ways of defining $\sin x$ and $\cos x$ are not by the elaborate geometrical methods of the previous chapter, but as the limits respectively of the series

$$x - \frac{x^3}{3!} + \frac{x^5}{5!} - \frac{x^7}{7!} + \text{etc.}\ldots,$$

and

$$1 - \frac{x^2}{2!} + \frac{x^4}{4!} - \frac{x^6}{6!} + \text{etc.}\ldots,$$

so that we put

$$\sin x = x - \frac{x^3}{3!} + \frac{x^5}{5!} - \frac{x^7}{7!} + \text{etc.}\ldots,$$

$$\cos x = 1 - \frac{x^2}{2!} + \frac{x^4}{4!} - \frac{x^6}{6!} + \text{etc.}\ldots.$$

These definitions are equivalent to the geometrical definitions, and both series can be proved to be convergent for all values of x, and uniformly convergent throughout any interval. These series for sine and cosine have a general likeness to the exponential series given above. They are, indeed, intimately connected with it by means of the theory of imaginary numbers explained in Chapters VII. and VIII.

The graph of the exponential function is given in fig. 29. It cuts the axis OY at the point $y = 1$, as evidently it ought to do, since when $x = 0$ every term of the series except the first is zero. The importance of the exponential function is that it represents any changing physical quantity whose rate of increase at any instant is a uniform percentage of its value

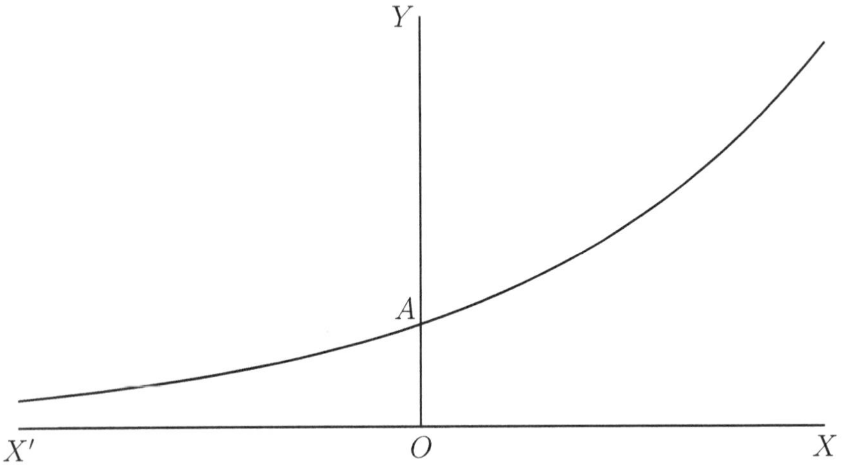

Fig. 29.

at that instant. For example, the above graph represents the size at any time of a population with a uniform birth-rate, a uniform death-rate, and no emigration, where the x corresponds to the time reckoned from any convenient day, and the y represents the population to the proper scale. The scale must be such that OA represents the population at the date which is taken as the origin. But we have here come upon the idea of "rates of increase" which is the topic for the next chapter.

An important function nearly allied to the exponential function is found by putting $-x^2$ for x as the argument in the exponential function. We thus get $\exp(-x^2)$. The graph $y = \exp(-x^2)$ is given in fig. 30.

The curve, which is something like a cocked hat, is called

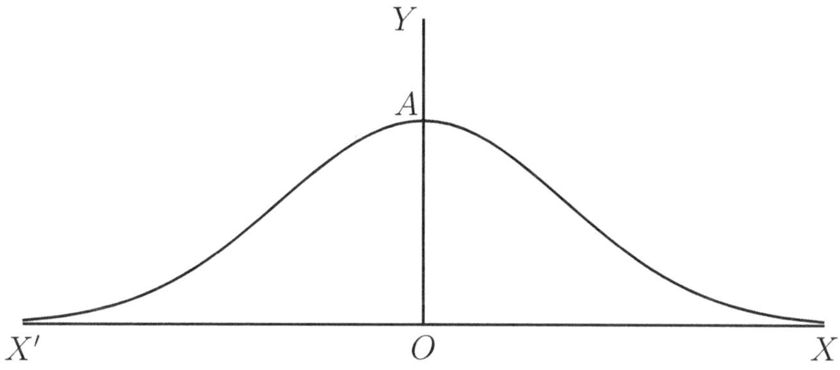

Fig. 30.

the curve of normal error. Its corresponding function is vitally important to the theory of statistics, and tells us in many cases the sort of deviations from the average results which we are to expect.

Another important function is found by combining the exponential function with the sine, in this way:—

$$y = \exp(-cx) \times \sin \frac{2\pi x}{p}.$$

Its graph is given in fig. 31. The points A, B, O, C, D, E, F, are placed at equal intervals $\frac{1}{2}p$, and an unending series of them should be drawn forwards and backwards. This function represents the dying away of vibrations under the influence of friction or of "damping" forces. Apart from the friction, the vibrations would be periodic, with a period p; but the influence of the friction makes the extent

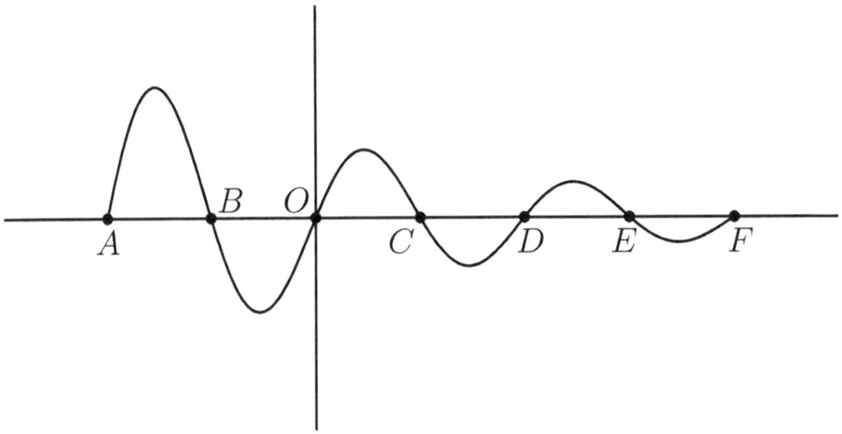

Fig. 31.

of each vibration smaller than that of the preceding by a constant percentage of that extent. This combination of the idea of "periodicity" (which requires the sine or cosine for its symbolism) and of "constant percentage" (which requires the exponential function for its symbolism) is the reason for the form of this function, namely, its form as a product of a sine-function into an exponential function.

CHAPTER XV

THE DIFFERENTIAL CALCULUS

THE invention of the differential calculus marks a crisis in the history of mathematics. The progress of science is divided between periods characterized by a slow accumulation of ideas and periods, when, owing to the new material for thought thus patiently collected, some genius by the invention of a new method or a new point of view, suddenly transforms the whole subject on to a higher level. These contrasted periods in the progress of the history of thought are compared by Shelley to the formation of an avalanche.

> The sun-awakened avalanche! whose mass,
> Thrice sifted by the storm, had gathered there
> Flake after flake,—in heaven-defying minds
> As thought by thought is piled, till some great truth
> Is loosened, and the nations echo round,
> ....

The comparison will bear some pressing. The final burst of sunshine which awakens the avalanche is not necessarily beyond comparison in magnitude with the other powers of nature which have presided over its slow formation. The same is true in science. The genius who has the good fortune to produce the final idea which transforms a whole region of thought, does not necessarily excel all his predecessors who have worked at the preliminary formation of ideas. In considering the history of science, it is both silly and ungrateful to confine our admiration with a gaping wonder to those men who have made the final advances towards a new epoch.

In the particular instance before us, the subject had a long history before it assumed its final form at the hands of its two inventors. There are some traces of its methods even among the Greek mathematicians, and finally, just before the actual production of the subject, Fermat (born 1601 A.D., and died 1665 A.D.), a distinguished French mathematician, had so improved on previous ideas that the subject was all but created by him. Fermat, also, may lay claim to be the joint inventor of coordinate geometry in company with his contemporary and countryman, Descartes. It was, in fact, Descartes from whom the world of science received the new ideas, but Fermat had certainly arrived at them independently.

We need not, however, stint our admiration either for Newton or for Leibniz. Newton was a mathematician and a student of physical science, Leibniz was a mathematician and a philosopher, and each of them in his own department of thought was one of the greatest men of genius that the world has known. The joint invention was the occasion of an unfortunate and not very creditable dispute. Newton was using the methods of Fluxions, as he called the subject, in 1666, and employed it in the composition of his *Principia*, although in the work as printed any special algebraic notation is avoided. But he did not print a direct statement of his method till 1693. Leibniz published his first statement in 1684. He was accused by Newton's friends of having got it from a MS. by Newton, which he had been shown privately. Leibniz also accused Newton of having plagiarized from him. There is now not very much doubt but that both should have the credit of being independent discoverers. The subject had arrived at a stage in which it was ripe for discovery, and there is nothing surprising in the fact that two such able men should have independently hit upon it.

These joint discoveries are quite common in science. Discoveries are not in general made before they have been led up to by the previous trend of thought, and by that time many minds are in hot pursuit of the important idea. If we merely keep to discoveries in which Englishmen are concerned, the simultaneous enunciation of the law of natural selection by Darwin and Wallace, and the simultaneous discovery of Neptune by Adams and the French astronomer, Leverrier, at once occur to the mind. The disputes, as to whom the credit ought to be given, are often influenced by an unworthy spirit of nationalism. The really inspiring reflection suggested by the history of mathematics is the unity of thought and interest among men of so many epochs, so many nations, and so many races. Indians, Egyptians, Assyrians, Greeks, Arabs, Italians, Frenchmen, Germans, Englishmen, and Russians, have all made essential contributions to the progress of the science. Assuredly the jealous exaltation of the contribution of one particular nation is not to show the larger spirit.

The importance of the differential calculus arises from the very nature of the subject, which is the systematic consideration of the rates of increase of functions. This idea is immediately presented to us by the study of nature; velocity is the rate of increase of the distance travelled, and acceleration is the rate of increase of velocity. Thus the fundamental idea of change, which is at the basis of our whole perception of phenomena, immediately suggests the enquiry as to the rate of change. The familiar terms of "quickly" and "slowly" gain their meaning from a tacit reference to rates of change. Thus the differential calculus is concerned with the very key of the position from which mathematics can be successfully applied to the explanation of the course of nature.

This idea of the rate of change was certainly in Newton's mind, and was embodied in the language in which he ex-

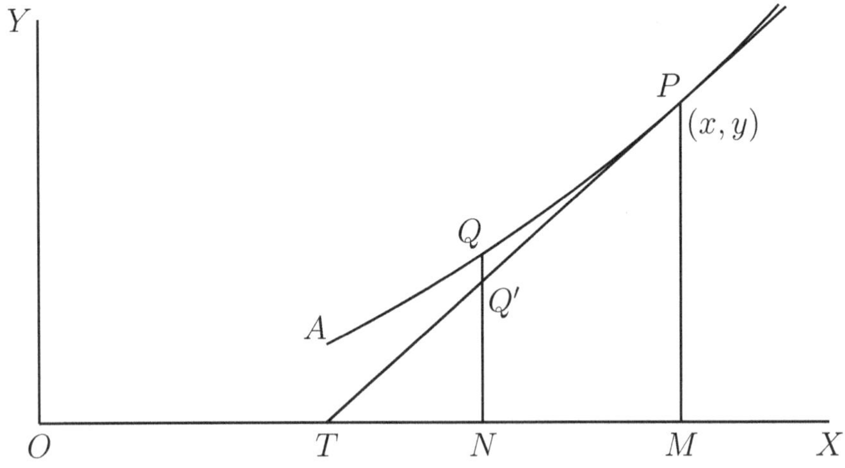

Fig. 32.

plained the subject. It may be doubted, however, whether

this point of view, derived from natural phenomena, was ever much in the minds of the preceding mathematicians who prepared the subject for its birth. They were concerned with the more abstract problems of drawing tangents to curves, of finding the lengths of curves, and of finding the areas enclosed by curves. The last two problems, of the rectification of curves and the quadrature of curves as they are named, belong to the Integral Calculus, which is however involved in the same general subject as the Differential Calculus.

The introduction of coordinate geometry makes the two points of view coalesce. For (*cf.* fig. 32) let AQP be any curved line and let PT be the tangent at the point P on it. Let the axes of coordinates be OX and OY; and let $y = f(x)$ be the equation to the curve, so that $OM = x$, and $PM = y$. Now let Q be any moving point on the curve, with coordinates x_1, y_1; then $y_1 = f(x_1)$. And let Q' be the point on the tangent with the same abscissa x_1; suppose that the coordinates of Q' are x_1 and y'. Now suppose that N moves along the axis OX from left to right with a uniform velocity; then it is easy to see that the ordinate y' of the point Q' on the tangent TP also increases uniformly as Q' moves along the tangent in a corresponding way. In fact it is easy to see that the ratio of the rate of increase of $Q'N$ to the rate of increase of ON is in the ratio of $Q'N$ to TN, which is the same at all points of the straight line. But the rate of increase of QN, which is the rate of increase of $f(x_1)$, varies from point to point of the curve so long as it is not straight. As Q passes through the point P, the rate of increase of $f(x_1)$

(where x_1 coincides with x for the moment) is the same as the rate of increase of y' on the tangent at P. Hence, if we have a general method of determining the rate of increase of a function $f(x)$ of a variable x, we can determine the slope of the tangent at any point (x, y) on a curve, and thence can draw it. Thus the problems of drawing tangents to a curve, and of determining the rates of increase of a function are really identical.

It will be noticed that, as in the cases of Conic Sections and Trigonometry, the more artificial of the two points of view is the one in which the subject took its rise. The really fundamental aspect of the science only rose into prominence comparatively late in the day. It is a well-founded historical generalization, that the last thing to be discovered in any science is what the science is really about. Men go on groping for centuries, guided merely by a dim instinct and a puzzled curiosity, till at last "some great truth is loosened."

Let us take some special cases in order to familiarize ourselves with the sort of ideas which we want to make precise. A train is in motion—how shall we determine its velocity at some instant, let us say, at noon? We can take an interval of five minutes which includes noon, and measure how far the train has gone in that period. Suppose we find it to be five miles, we may then conclude that the train was running at the rate of 60 miles per hour. But five miles is a long distance, and we cannot be sure that just at noon the train was moving at this pace. At noon it may have been running 70 miles per hour, and afterwards the brake may have been put on. It will be safer to work with a smaller interval, say one minute, which includes noon, and to measure the space traversed during that period. But for some purposes greater accuracy may be required, and one minute may be too long. In practice, the necessary inaccuracy of our measurements makes it useless to take too small a period for measurement. But in theory the smaller the period the better, and we are tempted to say that for ideal accuracy an infinitely small period is required. The older mathematicians, in particular Leibniz, were not only tempted, but yielded to the temptation, and did say it. Even now it is a useful fashion of speech, provided that we know how to interpret it into the language of common sense. It is curious that, in his exposition of the foundations of the calculus, Newton, the natural scientist, is much more philosophical than Leibniz, the philosopher, and on the other hand, Leibniz provided the admirable notation which has been so essential for the progress of the subject.

Now take another example within the region of pure mathematics. Let us proceed to find the rate of increase of the function x^2 for any value x of its argument. We have not yet really defined what we mean by rate of increase. We will try and grasp its meaning in relation to this particular case. When x increases to $x + h$, the function x^2 increases to $(x+h)^2$; so that the total increase has been $(x+h)^2 - x^2$, due to an increase h in the argument. Hence throughout the interval x to $(x+h)$ the average increase of the function per unit increase of the argument is $\dfrac{(x+h)^2 - x^2}{h}$. But

$$(x+h)^2 = x^2 + 2hx + h^2,$$

and therefore

$$\frac{(x+h)^2 - x^2}{h} = \frac{2hx + h^2}{h} = 2x + h.$$

Thus $2x + h$ is the average increase of the function x^2 per unit increase in the argument, the average being taken over by the interval x to $x + h$. But $2x + h$ depends on h, the size of the interval. We shall evidently get what we want, namely the *rate* of increase at the value x of the argument, by diminishing h more and more. Hence *in the limit* when h has *decreased indefinitely*, we say that $2x$ is the rate of increase of x^2 at the value x of the argument.

Here again we are apparently driven up against the idea of infinitely small quantities in the use of the words "in the limit when h has decreased indefinitely." Leibniz held that, mysterious as it may sound, there were actually existing such things as infinitely small quantities, and of course infinitely small numbers corresponding to them. Newton's language and ideas were more on the modern lines; but he did not succeed in explaining the matter with such explicitness so as to be evidently doing more than explain Leibniz's ideas in rather indirect language. The real explanation of the subject was first given by Weierstrass and the Berlin School of mathematicians about the middle of the nineteenth century. But between Leibniz and Weierstrass a copious literature, both mathematical and philosophical, had grown up round these mysterious infinitely small quantities which mathematics had discovered and philosophy proceeded to explain. Some philosophers, Bishop Berkeley, for instance, correctly denied the validity of the whole idea, though for reasons other than those indicated here. But the curious fact remained that, despite all criticisms of the foundations of the subject, there could be no doubt but that the mathematical procedure was substantially right. In fact, the subject was right, though the explanations were wrong. It is this possibility of being right, albeit with entirely wrong explanations as to what is being done, that so often makes external criticism—that is so far as it is meant to stop the pursuit of a method—singularly barren and futile in the progress of science. The instinct of trained observers, and their sense of curiosity, due to the fact that they are obviously getting at something, are far safer guides. Anyhow the general effect of the success of the Differential Calculus was to generate a large amount of bad philosophy, centring round the idea of the infinitely small. The relics of this verbiage may still be found in the explanations of many elementary mathematical text-books on the Differential Calculus. It is a safe rule to apply that, when a mathematical or philosophical author writes with a misty profundity, he is talking nonsense.

Newton would have phrased the question by saying that, as h approaches zero, in the limit $2x + h$ becomes $2x$. It is our task so to explain this statement as to show that it does not in reality covertly assume the existence of Leibniz's infinitely small quantities. In reading over the Newtonian method of statement, it is tempting to seek simplicity by saying that $2x + h$ is $2x$, when h is zero. But this will not do; for it thereby abolishes the interval from x to $x + h$, over which the average increase was calculated. The problem is, how to keep an interval of length h over which to calculate the average increase, and at the same time to treat h as if it were zero. Newton did this by the conception of a limit, and we now proceed to give Weierstrass's explanation of its real meaning.

In the first place notice that, in discussing $2x + h$, we have been considering x as fixed in value and h as varying. In other words x has been treated as a "constant" variable, or parameter, as explained in Chapter IX.; and we have really been considering $2x + h$ as a function of the argument h. Hence we can generalize the question on hand, and ask what we mean by saying that the function $f(h)$ tends to the limit l, say, as its argument h tends to the value zero. But again we shall see that the special value *zero* for the argument does not belong to the essence of the subject; and again we generalize still further, and ask, what we mean by saying that the function $f(h)$ tends to the limit l as h tends to the value a.

Now, according to the Weierstrassian explanation the whole idea of h tending to the value a, though it gives a sort of metaphorical picture of what we are driving at, is really off the point entirely. Indeed it is fairly obvious that, as long as we retain anything like "h tending to a," as a fundamental idea, we are really in the clutches of the infinitely small; for we imply the notion of h being infinitely near to a. This is just what we want to get rid of.

Accordingly, we shall yet again restate our phrase to be explained, and ask what we mean by saying that the limit of the function $f(h)$ at a is l.

The limit of $f(h)$ at a is a property of the neighbourhood of a, where "neighbourhood" is used in the sense defined in Chapter XI. during the discussion of the continuity of functions. The value of the function $f(h)$ at a is $f(a)$; but the limit is distinct in idea from the value, and may be different from it, and may exist when the value has not been defined. We shall also use the term "standard of approximation" in the sense in which it is defined in Chapter XI. In fact, in the definition of "continuity" given towards the end of that chapter we have practically defined a limit. The definition of a limit is:—

A function $f(x)$ has the limit l at a value a of its argument x, when in the neighbourhood of a its values approximate to l within *every* standard of approximation.

Compare this definition with that already given for continuity, namely:—

A function $f(x)$ is continuous at a value a of its argument, when in the neighbourhood of a its values approximate to its value at a within *every* standard of approximation.

It is at once evident that a function is continuous at a when (i) it possesses a limit at a, and (ii) that limit is equal to its value at a. Thus the illustrations of continuity which have been given at the end of Chapter XI. are illustrations of the idea of a limit, namely, they were all directed to proving that $f(a)$ was the limit of $f(x)$ at a for the functions considered and the value of a considered. It is really more instructive to consider the limit at a point where a function is not continuous. For example, consider the function of which the graph is given in fig. 20 of Chapter XI. This function $f(x)$ is defined to have the value 1 for all values of the argument except the integers 0, 1, 2, 3, etc., and for these integral values it has the value 0. Now let us think of its limit when $x = 3$. We notice that in the definition of the limit the value of the function at a (in this case, $a = 3$) is excluded. But, excluding $f(3)$, the values of $f(x)$, when x lies within any interval which (i) contains 3 not as an end-point, and (ii) does not extend so far as 2 and 4, are all equal to 1; and hence these values approximate to 1 within every standard of approximation. Hence 1 is the limit of $f(x)$ at the value 3 of the argument x, but by definition $f(3) = 0$.

This is an instance of a function which possesses both a value and a limit at the value 3 of the argument, but the value is not equal to the limit. At the end of Chapter XI. the function x^2 was considered at the value 2 of the argument.

Its value at 2 is 2^2, *i.e.* 4, and it was proved that its limit is also 4. Thus here we have a function with a value and a limit which are equal.

Finally we come to the case which is essentially important for our purposes, namely, to a function which possesses a limit, but no defined value at a certain value of its argument. We need not go far to look for such a function, $\frac{2x}{x}$ will serve our purpose. Now in any mathematical book, we might find the equation, $\frac{2x}{x} = 2$, written without hesitation or comment. But there is a difficulty in this; for when x is zero, $\frac{2x}{x} = \frac{0}{0}$; and $\frac{0}{0}$ has no defined meaning. Thus the value of the function $\frac{2x}{x}$ at $x = 0$ has no defined meaning. But for every other value of x, the value of the function $\frac{2x}{x}$ is 2. Thus the limit of $\frac{2x}{x}$ at $x = 0$ is 2, and it has no value at $x = 0$. Similarly the limit of $\frac{x^2}{x}$ at $x = a$ is a whatever a may be, so that the limit of $\frac{x^2}{x}$ at $x = 0$ is 0. But the value of $\frac{x^2}{x}$ at $x = 0$ takes the form $\frac{0}{0}$, which has no defined meaning. Thus the function $\frac{x^2}{x}$ has a limit but no value at 0.

We now come back to the problem from which we started this discussion on the nature of a limit. How are we going to define the rate of increase of the function x^2 at any value x of its argument. Our answer is that this rate of increase is the limit of the function $\frac{(x+h)^2 - x^2}{h}$ at the value zero for its argument h. (Note that x is here a "constant.") Let us see how this answer works in the light of our definition of a limit. We have

$$\frac{(x+h)^2 - x^2}{h} = \frac{2hx + h^2}{h} = \frac{h(2x+h)}{h}.$$

Now in finding the limit of $\frac{h(2x+h)}{h}$ at the value 0 of the argument h, the value (if any) of the function at $h = 0$ is excluded. But for all values of h, except $h = 0$, we can divide through by h. Thus the limit of $\frac{h(2x+h)}{h}$ at $h = 0$ is the same as that of $2x + h$ at $h = 0$. Now, whatever standard of approximation k we choose to take, by considering the interval from $-\frac{1}{2}k$ to $+\frac{1}{2}k$ we see that, for values of h which fall within it, $2x + h$ differs from $2x$ by less than $\frac{1}{2}k$, that is by less than k. This is true for *any* standard k. Hence in the neighbourhood of the value 0 for h, $2x+h$ approximates to $2x$ within *every* standard of approximation, and therefore $2x$ is the limit of $2x + h$ at $h = 0$. Hence by what has been said above $2x$ is the limit of $\frac{(x+h)^2 - x^2}{h}$ at the value 0 for h. It follows, therefore, that $2x$ is what we have called the rate of increase of x^2 at the value x of the argument. Thus this method conducts us to the same rate of increase for x^2 as did the Leibnizian way of making h grow "infinitely small."

The more abstract terms "differential coefficient," or "derived function," are generally used for what we have hitherto called the "rate of increase" of a function. The general definition is as follows: the differential coefficient of the function $f(x)$ is the limit, if it exist, of the function $\frac{f(x+h) - f(x)}{h}$ of the argument h at the value 0 of its argument.

How have we, by this definition and the subsidiary definition of a limit, really managed to avoid the notion of "infinitely small numbers" which so worried our mathematical forefathers? For them the difficulty arose because on the one hand they had to use an interval x to $x + h$ over which to calculate the average increase, and, on the other hand, they finally wanted to put $h = 0$. The result was they seemed to be landed into the notion of an existent interval of zero size. Now how do we avoid this difficulty? In this way—we use the notion that corresponding to *any* standard of approximation, *some* interval with such and such properties can be found. The difference is that we have grasped the importance of the notion of "the variable," and they had not done so. Thus, at the end of our exposition of the essential notions of mathematical analysis, we are led back to the ideas with which in Chapter II. we commenced our enquiry—that in mathematics the fundamentally important ideas are those of "*some* things" and "*any* things."

CHAPTER XVI

GEOMETRY

GEOMETRY, like the rest of mathematics, is abstract. In it the properties of the shapes and relative positions of things are studied. But we do not need to consider who is observing the things, or whether he becomes acquainted with them by sight or touch or hearing. In short, we ignore all particular sensations. Furthermore, particular things such as the Houses of Parliament, or the terrestrial globe are ignored. Every proposition refers to any things with such and such geometrical properties. Of course it helps our imagination to look at particular examples of spheres and cones and triangles and squares. But the propositions do not merely apply to the actual figures printed in the book, but to any such figures.

Thus geometry, like algebra, is dominated by the ideas of "any" and "some" things. Also, in the same way it studies the interrelations of sets of things. For example, consider any two triangles ABC and DEF.

What relations must exist between some of the parts of these triangles, in order that the triangles may be in all respects equal? This is one of the first investigations undertaken in all elementary geometries. It is a study of a certain set of possible correlations between the two triangles. The answer is that the triangles are in all respects equal, if:— Either, (a) Two sides of the one and the included angle are

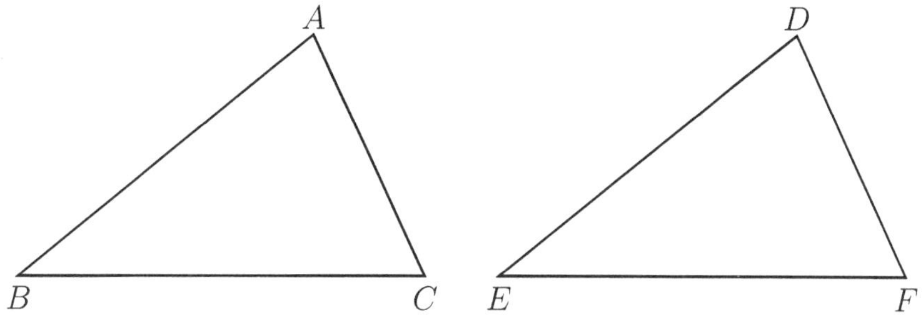

Fig. 33.

respectively equal to two sides of the other and the included angle:

Or, (b) Two angles of the one and the side joining them are respectively equal to two angles of the other and the side joining them:

Or, (c) Three sides of the one are respectively equal to three sides of the other.

This answer at once suggests a further enquiry. What is the nature of the correlation between the triangles, when the three angles of the one are respectively equal to the three angles of the other? This further investigation leads us on to the whole theory of similarity (*cf.* Chapter XIII.), which is another type of correlation.

Again, to take another example, consider the internal structure of the triangle ABC. Its sides and angles are inter-related—the greater angle is opposite to the greater side, and the base angles of an isosceles triangle are equal. If we proceed to trigonometry this correlation receives a more exact determination in the familiar shape

$$\frac{\sin A}{a} = \frac{\sin B}{b} = \frac{\sin C}{c},$$

$a^2 = b^2 + c^2 - 2bc \cos A$, with two similar formulæ.

Also there is the still simpler correlation between the angles of the triangle, namely, that their sum is equal to two right angles; and between the three sides, namely, that the sum of the lengths of any two is greater than the length of the third.

Thus the true method to study geometry is to think of interesting simple figures, such as the triangle, the parallelogram, and the circle, and to investigate the correlations between their various parts. The geometer has in his mind not a detached proposition, but a figure with its various parts mutually inter-dependent. Just as in algebra, he generalizes the triangle into the polygon, and the side into the conic section. Or, pursuing a converse route, he classifies triangles according as they are equilateral, isosceles, or scalene, and polygons according to their number of sides, and conic sections according as they are hyperbolas, ellipses, or parabolas.

The preceding examples illustrate how the fundamental ideas of geometry are exactly the same as those of algebra; except that algebra deals with numbers and geometry with lines, angles, areas, and other geometrical entities. This fundamental identity is one of the reasons why so many geometrical truths can be put into an algebraic dress. Thus if A, B, and C are the numbers of degrees respectively in the angles of the triangle ABC, the correlation between the angles is represented by the equation

$$A + B + C = 180°;$$

and if a, b, c are the number of feet respectively in the three sides, the correlation between the sides is represented by $a < b + c$, $b < c + a$, $c < a + b$. Also the trigonometrical formulæ quoted above are other examples of the same fact. Thus the notion of the variable and the correlation of variables is just as essential in geometry as it is in algebra.

But the parallelism between geometry and algebra can be pushed still further, owing to the fact that lengths, areas, volumes, and angles are all measurable; so that, for example, the size of any length can be determined by the number (not necessarily integral) of times which it contains some arbitrarily known unit, and similarly for areas, volumes, and angles. The trigonometrical formulæ, given above, are examples of this fact. But it receives its crowning application in analytical geometry. This great subject is often misnamed as Analytical Conic Sections, thereby fixing attention on merely one of its subdivisions. It is as though the great science of Anthropology were named the Study of Noses, owing to the fact that noses are a prominent part of the human body.

Though the mathematical procedures in geometry and algebra are in essence identical and intertwined in their development, there is necessarily a fundamental distinction between the properties of space and the properties of number—in fact all the essential difference between space and number.

The "spaciness" of space and the "numerosity" of number are essentially different things, and must be directly apprehended. None of the applications of algebra to geometry or of geometry to algebra go any step on the road to obliterate this vital distinction.

One very marked difference between space and number is that the former seems to be so much less abstract and fundamental than the latter. The number of the archangels can be counted just because they are things. When we once know that their names are Raphael, Gabriel, and Michael, and that these distinct names represent distinct beings, we know without further question that there are three of them. All the subtleties in the world about the nature of angelic existences cannot alter this fact, granting the premisses.

But we are still quite in the dark as to their relation to space. Do they exist in space at all? Perhaps it is equally nonsense to say that they are here, or there, or anywhere, or everywhere. Their existence may simply have no relation to localities in space. Accordingly, while numbers must apply to all things, space need not do so.

The perception of the locality of things would appear to accompany, or be involved in many, or all, of our sensations. It is independent of any particular sensation in the sense that it accompanies many sensations. But it is a special peculiarity of the things which we apprehend by our sensations. The direct apprehension of what we mean by the positions of things in respect to each other is a thing *sui generis*, just as are the apprehensions of sounds, colours, tastes, and smells.

At first sight therefore it would appear that mathematics, in so far as it includes geometry in its scope, is not abstract in the sense in which abstractness is ascribed to it in Chapter I.

This, however, is a mistake; the truth being that the "spaciness" of space does not enter into our geometrical *reasoning* at all. It enters into the geometrical intuitions of mathematicians in ways personal and peculiar to each individual. But what enter into the reasoning are merely certain properties of things in space, or of things forming space, which properties are completely abstract in the sense in which abstract was defined in Chapter I.; these properties do not involve any peculiar space-apprehension or space-intuition or space-sensation. They are on exactly the same basis as the mathematical properties of number. Thus the space-intuition which is so essential an aid to the study of geometry is logically irrelevant: it does not enter into the premisses when they are properly stated, nor into any step of the reasoning. It has the practical importance of an example, which is essential for the stimulation of our thoughts. Examples are equally necessary to stimulate our thoughts on number. When we think of "two" and "three" we see strokes in a row, or balls in a heap, or some other physical aggregation of particular things. The peculiarity of geometry is the fixity and overwhelming importance of the one particular example which occurs to our minds. The abstract logical form of the propositions when fully stated is, "If any collections of things have such and such abstract properties, they also have such and such other abstract properties." But what appears before the mind's eye is a collection of points, lines, surfaces, and volumes in the space: this example inevitably appears, and is the sole example which lends to the proposition its interest. However, for all its overwhelming importance, it is but an example.

Geometry, viewed as a mathematical science, is a division of the more general science of order. It may be called the science of dimensional order; the qualification "dimensional" has been introduced because the limitations, which reduce it to only a part of the general science of order, are such as to produce the regular relations of straight lines to planes, and of planes to the whole of space.

It is easy to understand the practical importance of space in the formation of the scientific conception of an external physical world. On the one hand our space-perceptions are intertwined in our various sensations and connect them together. We normally judge that we touch an object in the same place as we see it; and even in abnormal cases we touch it in the same space as we see it, and this is the real fundamental fact which ties together our various sensations. Accordingly, the space perceptions are in a sense the common part of our sensations. Again it happens that the abstract properties of space form a large part of whatever is of spatial interest. It is not too much to say that to every property of space there corresponds an abstract mathematical statement. To take the most unfavourable instance, a curve may have a special beauty of shape: but to this shape there will correspond some abstract mathematical properties which go with this shape and no others.

Thus to sum up: (1) the properties of space which are investigated in geometry, like those of number, are properties belonging to things as things, and without special reference to any particular mode of apprehension: (2) Space-perception accompanies our sensations, perhaps all of them, certainly many; but it does not seem to be a necessary quality of things that they should all exist in one space or in any space.

CHAPTER XVII

QUANTITY

In the previous chapter we pointed out that lengths are measurable in terms of some unit length, areas in term of a unit area, and volumes in terms of a unit volume.

When we have a set of things such as lengths which are measurable in terms of any one of them, we say that they are quantities of the same kind. Thus lengths are quantities of the same kind, so are areas, and so are volumes. But an area is not a quantity of the same kind as a length, nor is it of the same kind as a volume. Let us think a little more on what is meant by being measurable, taking lengths as an example.

Lengths are measured by the foot-rule. By transporting the foot-rule from place to place we judge of the equality of lengths. Again, three adjacent lengths, each of one foot, form one whole length of three feet. Thus to measure lengths we have to determine the equality of lengths and the addition of lengths. When some test has been applied, such as the transporting of a foot-rule, we say that the lengths are equal; and when some process has been applied, so as to secure lengths being contiguous and not overlapping, we say that the lengths have been added to form one whole length. But we cannot arbitrarily take any test as the test of equality and any process as the process of addition. The results of operations of addition and of judgments of equality must be in accordance with certain preconceived conditions. For example, the addition of two greater lengths must yield a length greater than that yielded by the addition of two smaller lengths. These preconceived conditions when accurately formulated may be called axioms of quantity. The only question as to their truth or falsehood which can arise is whether, when the axioms are satisfied, we necessarily get what ordinary people call quantities. If we do not, then the name "axioms of quantity" is ill-judged—that is all.

These axioms of quantity are entirely abstract, just as are the mathematical properties of space. They are the same for all quantities, and they presuppose no special mode of perception. The ideas associated with the notion of quantity are the means by which a continuum like a line, an area, or a volume can be split up into definite parts. Then these parts are counted; so that numbers can be used to determine the exact properties of a continuous whole.

Our perception of the flow of time and of the succession of events is a chief example of the application of these ideas of quantity. We measure time (as has been said in considering periodicity) by the repetition of similar events—the burning of successive inches of a uniform candle, the rotation of the earth relatively to the fixed stars, the rotation of the hands of a clock are all examples of such repetitions. Events of these types take the place of the foot-rule in relation to lengths. It is not necessary to assume that events of any one of these types are exactly equal in duration at each recurrence. What is necessary is that a rule should be known which will enable us to express the relative durations of, say, two examples of some type. For example, we may if we like suppose that the rate of the earth's rotation is decreasing, so that each day is longer than the preceding by some minute fraction of a second. Such a rule enables us to compare the length of any day with that of any other day. But what is essential is that one series of repetitions, such as successive days, should be taken as the standard series; and, if the various events of that series are not taken as of equal duration, that a rule should be stated which regulates the duration to be assigned to each day in terms of the duration of any other day.

What then are the requisites which such a rule ought to have? In the first place it should lead to the assignment of nearly equal durations to events which common sense judges to possess equal durations. A rule which made days of violently different lengths, and which made the speeds of apparently similar operations vary utterly out of proportion to the apparent minuteness of their differences, would never do. Hence the first requisite is general agreement with common sense. But this is not sufficient absolutely to determine the rule, for common sense is a rough observer and very easily satisfied. The next requisite is that minute adjustments of the rule should be so made as to allow of the simplest possible statements of the laws of nature. For example, astronomers tell us that the earth's rotation is slowing down, so that each day gains in length by some inconceivably minute fraction of a second. Their only reason for their assertion (as stated more fully in the discussion of periodicity) is that without it they would have to abandon the Newtonian laws of motion. In order to keep the laws of motion simple, they alter the measure of time. This is a perfectly legitimate procedure so long as it is thoroughly understood.

What has been said above about the abstract nature of the mathematical properties of space applies with appropriate verbal changes to the mathematical properties of time. A sense of the flux of time accompanies all our sensations and perceptions, and practically all that interests us in regard to time can be paralleled by the abstract mathematical properties which we ascribe to it. Conversely what has been said about the two requisites for the rule by which we determine the length of the day, also applies to the rule for determining the length of a yard measure—namely, the yard measure appears to retain the same length as it moves about. Accordingly, any rule must bring out that, apart from minute changes, it does remain of invariable length; Again, the second requisite is this, a definite rule for minute changes shall be stated which allows of the simplest expression of the laws of nature. For example, in accordance with the second requisite the yard measures are supposed to expand and contract with changes of temperature according to the substances which they are made of.

Apart from the facts that our sensations are accompanied with perceptions of locality and of duration, and that lines, areas, volumes, and durations, are each in their way quantities, the theory of numbers would be of very subordinate use in the exploration of the laws of the Universe, As it is, physical science reposes on the main ideas of number, quantity, space, and time. The mathematical sciences associated with them do not form the whole of mathematics, but they are the substratum of mathematical physics as at present existing.

NOTES

A (p. 45).—In reading these equations it must be noted that a bracket is used in mathematical symbolism to mean that the operations within it are to be performed first. Thus $(1 + 3) + 2$ directs us first to add 3 to 1, and then to add 2 to the result; and $1 + (3 + 2)$ directs us first to add 2 to 3, and then to add the result to 1. Again a numerical example of equation (5) is

$$2 \times (3 + 4) = (2 \times 3) + (2 \times 4).$$

We perform first the operations in brackets and obtain

$$2 \times 7 = 6 + 8$$

which is obviously true.

B (p. 110).—This fundamental ratio $\frac{SP}{PN}$ is called the eccentricity of the curve. The shape of the curve, as distinct from its scale or size, depends upon the value of its eccentricity. Thus it is wrong to think of ellipses in general or of hyperbolas in general as having in either case one definite shape. Ellipses with different eccentricities have different shapes, and their sizes depend upon the lengths of their major axes. An ellipse with small eccentricity is very nearly a circle, and an ellipse of eccentricity only slightly less than unity is a long flat oval. All parabolas have the same eccentricity and are therefore of the same shape, though they can be drawn to different scales.

C (p. 168).—If a series with all its terms positive is convergent, the modified series found by making some terms positive and some negative according to any definite rule is also convergent. Each one of the set of series thus found, including the original series, is called "absolutely convergent." But it is possible for a series with terms partly positive and partly negative to be convergent, although the corresponding series with all its terms positive is divergent. For example, the series

$$1 - \tfrac{1}{2} + \tfrac{1}{3} - \tfrac{1}{4} + \text{etc.}$$

is convergent though we have just proved that

$$1 + \tfrac{1}{2} + \tfrac{1}{3} + \tfrac{1}{4} + \text{etc.}$$

is divergent. Such convergent series, which are not absolutely convergent, are much more difficult to deal with than absolutely convergent series.

BIBLIOGRAPHY

NOTE ON THE STUDY OF MATHEMATICS

THE difficulty that beginners find in the study of this science is due to the large amount of technical detail which has been allowed to accumulate in the elementary text-books, obscuring the important ideas.

The first subjects of study, apart from a knowledge of arithmetic which is presupposed, must be elementary geometry and elementary algebra. The courses in both subjects should be short, giving only the necessary ideas; the algebra should be studied graphically, so that in practice the ideas of elementary coordinate geometry are also being assimilated. The next pair of subjects should be elementary trigonometry and the coordinate geometry of the straight line and circle. The latter subject is a short one; for it really merges into the algebra. The student is then prepared to enter upon conic sections, a very short course of geometrical conic sections and a longer one of analytical conics. But in all these courses great care should be taken not to overload the mind with more detail than is necessary for the exemplification of the fundamental ideas.

The differential calculus and afterwards the integral calculus now remain to be attacked on the same system. A good teacher will already have illustrated them by the consideration of special cases in the course on algebra and coordinate geometry. Some short book on three-dimensional geometry must be also read.

This elementary course of mathematics is sufficient for some types of professional career. It is also the necessary preliminary for any one wishing to study the subject for its intrinsic interest. He is now prepared to commence on a more extended course. He must not, however, hope to be able to master it as a whole. The science has grown to such vast proportions that probably no living mathematician can claim to have achieved this.

Passing to the serious treatises on the subject to be read *after* this preliminary course, the following may be mentioned: Cremona's *Pure Geometry* (English Translation, Clarendon Press, Oxford), Hobson's *Treatise on Trigonometry*, Chrystal's *Treatise on Algebra* (2 volumes), Salmon's *Conic Sections*, Lamb's *Differential Calculus*, and some book on *Differential Equations*. The student will probably not desire to direct equal attention to all these subjects, but will study one or more of them, according as his interest dictates. He will then be prepared to select more advanced works for himself, and to plunge into the higher parts of the subject. If his interest lies in analysis, he should now master an elementary treatise on the theory of Functions of the Complex Variable; if he prefers to specialize in Geometry, he must now proceed to the standard treatises on the Analytical Geometry of three dimensions. But at this stage of his career in learning he will not require the advice of this note.

I have deliberately refrained from mentioning any elementary works. They are very numerous, and of various merits, but none of such outstanding superiority as to require special mention by name to the exclusion of all the others.

INDEX

This book has been produced in line with the EU GPSR guidelines about the safety of products.

The General Product Safety Regulation is the European Union's updated framework for ensuring that all consumer products, including books, are safe for consumers.

This book has been printed by CPI books GmbH. The printer has issued safety certificates for the materials - like ink, paper and glue - being used.

The product identifier is: 9786256629622

The author is responsible for the content of the book, is the publisher of the works and bears full responsibility for it.

The book has been published via Bookmundo. Bookmundo enables any author to share their stories with the rest of the world via printed books and ebooks and a broad distribution network.

Bookmundo will act as an intermediary in regard to questions about safety and will address them to the printer / author. Should there be any question in regard to the safety of the product, please contact us.

Bookmundo
Delftsestraat 33
3013AE Rotterdam
The Netherlands
info@bookmundo.com

FSC
www.fsc.org
MIX
Papier | Fördert
gute Waldnutzung
FSC® C083411

Kolibri 360 GmbH
Ferdinand-Jühlke-Straße 7
99095 Erfurt, Deutschland
produktsicherheit@kolibri360.de